The Worm Cafe

Flower Press

For my mother's mother Lucille, my greatest teacher.

BP

The Worm Cafe:

Mid-Scale Vermicomposting of Lunchroom Wastes —

A Manual for Schools, Small Businesses, and Community Groups

by
Binet Payne
with illustrations by Paul E. Bourgeois

Flower Press
Kalamazoo, Michigan USA

"Variety of garbage eaten by worms," artist, Mary Frances Fenton, adapted from "Kinds of garbage eaten by worms" in *Winter Composting with Worms*, final report from the Kalamazoo Nature Center to the National Center for Appropriate Technology, 1979.

"Worm-food waste ratio," artist, Mary Frances Fenton. *Worms Eat My Garbage*, © 1982, 1997.

The Worm Cafe: Mid-Scale Vermicomposting of Lunchroom Waste — A Manual for Schools, Small Businesses, and Community Groups, by Binet Payne.

Publisher's Cataloging-in-Publication
(*Provided by Quality Books, Inc.*)

Payne, Binet.
The worm cafe : mid-scale vermicomposting of lunchroom wastes : a manual for schools, small businesses, and community groups / by Binet Payne ; with illustrations by Paul Bourgeois ; [edited by Diane D. Worden]. — 1st ed. — Kalamazoo, Mich. : Flower Press, 1999.
p. cm.
Includes bibliographical references and index.
ISBN: 0-942256-11-5

1. Compost. 2. Earthworm culture. 3. Schools—Waste disposal. 4. Small business—Waste disposal. 5. Ecology—Study and teaching (Elementary) I. Title.

S661.P39 1999 631.8'75
QBI98-1627

Flower Press
10332 Shaver Road
Kalamazoo, Michigan 49024

Contents

List of Figures

Preface

My reasons for writing this book are to inspire, inform, and encourage school districts to undertake mid-scale vermicomposting projects. The book is also a wake-up call to parents, school boards, administrators, and teachers to take a closer look at the "hidden" curriculum taught in our schools. Research indicates that the most powerful tool in teaching is modeling the behavior or skill desired in students. What are we demonstrating to our children with practices which promote fast food in our cafeterias, disposable trays, utensils, and cups? If our intent is to truly teach them responsible stewardship of the Earth, then we must be willing to make radical changes to the status quo.

If we want to guarantee the survival of our species and our planet, we must educate the leaders of tomorrow — our children. School practices which are not ecologically sound should be challenged and changed. Mahatma Gandhi reportedly once said that if we wanted to change the world, we must become the change we would wish to see in the world. My message to readers is to help them be the change. Work with your school or place of business to stop mixing food waste with paper and burying it into landfills where it sits forever.

Today, the development of Earth-centered values is crucial to the survival of all species, including our own. If we are going to effect change, we need to begin with the attitudes and beliefs of our children. They need to understand the interconnectedness of all life, and how it relates to them and to the world around them. In classrooms today, we have a tendency to present students with the problems, yet give them no tools to reach a solution or even to develop a plan. When it comes to the environment, we need to give our children viable plans and solutions — solutions which sustain and nourish them and the planet we all share.

This publication describes how to set up, maintain, and manage a mid-scale vermicomposting program at a school or a place of business, and shows that vermicomposting, the processing of organic wastes by redworms, is practical. Mid-scale vermicomposting is defined as larger than a home system which processes from two to six pounds (<1 to <3kg) of food waste per week, yet smaller than commercial operations which process tons of waste per day. Each mid-scale system I describe handles about 25 pounds (>11kg) of food waste per week. Whenever weight, length, or volume measurements are important in this manual, metric units in parentheses follow the English system of pounds, feet, and gallons used in the United States. All metrics have been rounded up or down to the nearest half unit.

I hope readers use this book as a practical tool to change the ways schools or businesses handle their solid waste disposal. With the students of Laytonville Middle School, whose help and support underlie this book, I call for change. Together we believe that convincing other schools to vermicompost their food and paper wastes will make a real difference in the health of their communities. If, by example, we help divert tons and tons of paper and garbage from landfills all over the world, creating rich organic compost at the same time, we will make a real difference. The world will be a much better place. I believe that much of the alienation today's young adults feel comes from being disconnected from the natural world. Vermicomposting and worm science is one way to reconnect our children to the natural world and help them find their place in that order.

In a school-wide vermicomposting program, another more basic goal should be the creation of a "systemic system," meaning one that permeates all levels of the organization. It must not depend on any one individual for its success. Take the time to get the cooperation of your support staff, teachers, administration, and most importantly, the children. Because the program at Laytonville Elementary and Middle Schools is managed by its students, the size of all tools, containers, and everything connected to the project is deliberately comfortable for them to handle.

I learned about the benefits of composting and gardening from my grandmother, Lucille Ruppert, born May 31, 1897.

Her family brewed Ruppert beer. She was a native of San Francisco who spent her childhood on the family ranch in Half Moon Bay. Family outings often meant taking the stagecoach into the city to attend the opera and plays. Later she and her family moved into San Francisco, but I suspect she always felt more at home on the family ranch. When she came of age, she married a poor young farmer, Mr. Even Morlan, who had followed his family from Iowa to California in a covered wagon with two teams of horses and one cow. My grandfather first labored in the oil fields of southern California. Together, the new couple purchased land near the railroad, and they waited for the town of La Habra to spring up nearby.

My earliest remembrances as a child took place on that small farm. It was there that I learned from my grandmother, whom I called Nana, what I consider to be the most valuable lessons of my life.

In the summertime while my mother worked, I stayed with Nana. I would arrive at her house freshly scrubbed, with my hair in ringlets and dressed in petticoats and lace. Once inside the house, I quickly changed into my coveralls and flannel shirt. Together we became co-conspirators in my quest to get as dirty as possible in the eight hours I was left in her charge. We fed chickens, made compost, robbed honey from her hives, planted vegetables, gathered eggs, and picked tomatoes still warm from the summer sun. My Nana sold Lollo berries (a type of blackberry), eggs, and vegetables from a stand at the edge of the road which ran in front of her house. By one o'clock each day, I fell asleep from exhaustion atop her feather bed. And each day, just before my mother returned to pick me up, I would again be scrubbed squeaky clean and dressed just as my mother had left me with every single ringlet in place.

My grandmother nurtured in me an appreciation for the beauty, strength, and diversity one finds in the garden. Through her teachings, I learned a deep-rooted connectedness with the world around me. In her garden I began to understand the cycle of life: birth, life, growth, and eventual death with its decay and rebirth. I witnessed the link of life from the garden to the table to the compost bucket. I learned that for every action there is an equal reaction, and that we are all connected in the web of life.

In writing this book, I focused on mid-scale vermicomposting and how it can be used effectively in a school setting. I would be remiss not to commend my mentor Mary Appelhof and her books, *Worms Eat My Garbage* (Flower Press, 1982, 1997), which describe how to set up and maintain a worm composting system at home.

Currently I use *Worms Eat My Garbage* as a textbook for my 6th-, 7th-, and 8th-grade students when teaching earthworm science. This book is extremely user-friendly. It gives in-depth answers to many scientific questions about the biological needs of earthworms and home vermicomposting systems. Vermicomposting bins, though easy to maintain, are very complicated ecological systems. If you are curious and want to learn more about earthworms, please consult Appelhof's book.

For teachers, I also recommend her second book, *Worms Eat Our Garbage: Classroom Activities for a Better Environment*, co-authored with Mary Frances Fenton and Barbara Loss Harris. Designed for grades four through eight, this curriculum guide is packed with lessons and activities which really support instruction in the classroom.

Acknowledgments

Mary Appelhof is a scientist, educator, and author with master's degrees in biology and education, who also operates the publishing and mail order catalog company, Flowerfield Enterprises. Her early work with home vermicomposting was the original inspiration for my school experimentation that has become successful mid-scale vermicomposting. Without Appelhof's generous help, encouragement, and direct financial support, this book would not be a reality today.

Many other people contributed to the completion of this book. I want to thank Dr. Roy Hartenstein, retired SUNY-Syracuse, for the research information he provided. His laboratory work was very valuable to me as I tried to calculate how long it would take for our bins to reach optimum worm population. For reviewing Chapter 6 my thanks go to Drs. Jill Clapperton, Lethbridge Research Centre, and Sam James, Maharishi University of Management, taxonomic scientists;

and for Chapters 7 and 11, Dr. Elaine Ingham, an Oregon State University expert on the soil foodweb.

I also want to thank the staff of Laytonville Elementary and Middle Schools, Mendocino County, California, who believed in and helped to fund the project when budgets were often tight. High school students Brian Bowles and Sky Cassidy provided valuable aid and helped compile data. Above all, the students at Laytonville Middle School gave countless hours of care and attention to the worms and the project. Rebecca Norris, Meghan Champion, Katelyn Rogoff, Ashley Smith, and Amber Utford graciously shared their computer-assisted assignments with me for this book.

My thanks go as well to all the schools and teachers and educators elsewhere who have shared information with me as they have embarked on similar projects. Thank you Marianne Loeser, language arts teacher and friend, for helping lend credibility to the project. She believed in the importance of project-based learning before it had a name, and helped me convince a skeptical principal, Richard Matlock, that worms were the future in school-waste recycling. Thanks for taking the risk, Richard. Thank you, also, Pam Cornell and Bonnie Frost, the two leaders in the kitchen, for being supportive and helpful; without them this project would have never happened.

Not to forget the woman who gave me the original bucket of manure worms, my gratitude also goes to Adele Buck. Little did we realize at the time what that bucket of worms would create.

I want to thank Harwood Products, its owners the Harwood family, friends, and employees. Their support in the early years of this project helped us through the lean years of school budget cuts. They supplied lumber for our endeavors to support a project they believed to be valuable for the children in our community.

Thanks to both the community of Laytonville and its School Site Council for the support given my projects, ranging from fundraisers to volunteer work days. Thank you, Albert Repovsch, as friend of the garden and my greatest volunteer. I also thank Ginger Firks who grew up with me in the sixties, "What a long strange trip its been . . . ." As my dear friend, she gave me the time I needed to complete this project.

I want to thank Mary Joens Poulton and Katie Joens Poulton for their work on the mailing list and the annotated bibliography. They saved me many hours of work. Special thanks to my neighbor and artist, Paul Bourgeois, for his original and sensitive illustrations; his creativity brought life to these pages. Thank you, Paul. In contrast to his drawings, photographs used in this book were taken by Mary Appelhof, my husband Dan, and myself. Eric Fritz prepared these photos and the anatomy graphics labels for publication. Then I want to thank Nancy Essex at Flower Press for picking up the phone and saying, "I am taking over this project, let's get it done!" I am sure that without her prompting this project would not be complete. I also want to thank Diane Worden, who converted a muddled mess of words into a finished manuscript, and the final proofreader, Maria Perez-Stable.

Thank you, Bea and Judd Payne, my husband's mother and father for their deep connection with the Earth and practical wisdom that has taught me much about life. Especially, I thank my loving husband, Dannie, for his love, support, and patience. As my center and inspiration, he wholeheartedly arranged his time so I could complete this project. Our time together is coming; may living with me always be an adventure. Someday I will decide what I want to be when I grow up, I promise!

And finally, I want to thank all my family who shared their best selves with me. My parents, Mary Ellen and Freeman Rogers, respectively, gifted me with a fine sense of tradition and a way with words. My dear sister Linda believes there is good in everyone of us and searches until she finds it, and my brother Deak shares the mystery and wonder of nature with me. Love, encouragement, and childhood wonderment from my two daughters, Jennifer and Maya, taught me so much.

David W. Orr wrote, "All education is environmental education. By what is included or excluded, students are taught that they are part of or apart from the natural world." May this book support teachers everywhere in their attempts to convince their schools or favorite place of business that mid-scale vermicomposting is a safe, efficient, and sustainable way to reduce solid waste, water, and fossil fuel costs. Lastly, may the work of Laytonville's students serve as a model for other schools across the nation.

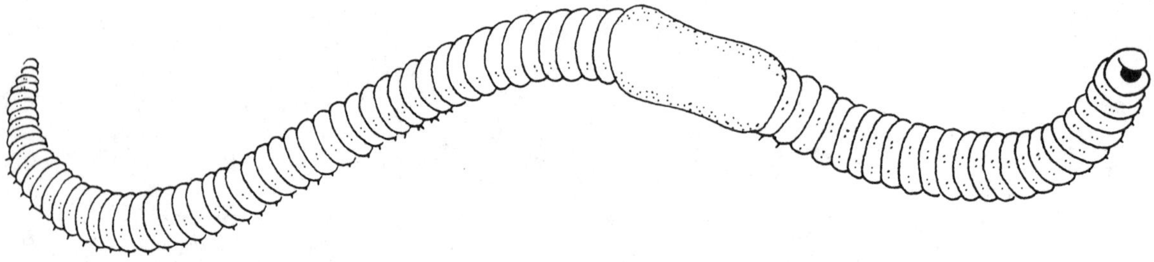

Introduction

Today's students are hungry for work that is real, for learning that is meaningful. Project-based learning is a teaching strategy that allows students to take more responsibility for their learning as they make decisions and create solutions to problems that interest them. All subjects can be integrated as students apply their academic, social, and life skills to their work. Project-based learning is a teaching strategy that honors students as capable people, readying themselves to be future leaders by giving them chances to be leaders as children.

Project-based learning encourages schools to become hubs of community connections rather than isolated organizations. Many schools focused on project-based learning have joined the PBL Network created by the Autodesk Foundation. At the Autodesk Foundation, our primary focus is the promotion and support of project-based learning. We see project-based learning as the most effective strategy for enabling students to have successful lives, as children and adults. We appreciate the way project-based learning provides students with the skills they need to succeed: lifelong learning, critical thinking, collaborative skills, creativity, organizational skills, and systems thinking. We recognize that schools focused on project-based learning create opportunities for students to highlight their strengths and challenge their weaker areas.

Interdependence is evident as students work together with each other, other classes, and community to achieve their goals. Because of its holistic approach centered on relationships, project-based learning is especially well-suited to enable students to learn about the environment. With project-based learning, students see that diversity is strength, every connection has its effect, and they are a valued part of the whole. Environmental projects such as school gardens, habitat restoration, water-monitoring, and recycling enable students to make a difference, to participate in natural cycles in a positive way. "Worm science," including vermiculture, is an especially effective way to provide students with first-hand experience with an important life cycle.

Through her example and many presentations, Binet Payne has inspired many teachers to allow their students to "learn by doing," constructing and implementing their own garden/composting projects. Worm science is an excellent hands-on experience that integrates all subjects as it teaches ecoliteracy. In this book, Binet has detailed how to facilitate your own mid-scale worm composting program through project-based learning. The process she details gives ownership to the students, allowing them to do all that they can. As you and your students get started with your own worm science activities, know that the "community you build in the bin" will foster community in your classroom and beyond.

Laurette Rogers
Autodesk Foundation
June 18, 1998

Chapter 1

The New Three Rs

School curriculums have included readin', 'riting, and 'rithmetic for hundreds of years. Today, population growth and changes in society have led to more people and fewer natural resources. It now becomes important to heed new educational imperatives. Reduce, reuse, and recycle are major economic corollaries of vermicomposting. As schools change, so will our impact on the future. School vermicomposting projects and school gardens, begun by Laytonville Middle School in the fall of 1987, can demonstrate that communities can reduce their use of local landfills, water, and fossil fuels.

In 1989 only about five percent of the produce grown in the U.S. was organically grown. Experts project that, with proper incentives, this figure could easily increase to ten percent in the year 2000. To help effectively reduce world hunger and reduce the harmful toxic effects of an industrialized society, we must develop safer methods of agriculture which sustain the Earth and its resources. To do this, we must reduce our use of fossil fuels and pesticides, and conserve topsoil and water.

When fertilizers were found in drinking water samples of Whitehall, Illinois, by the Environmental Working Group, a part of the National Research Council, EWG's Brian Cohen estimated that, "every river and reservoir in the Midwest is contaminated with corn fertilizers." Further, according to *Newsweek* (June 1998), they also warned us that current law does not adequately protect children from pesticides. Government data analyzed in 1998 show that over 1,000,000 children consume pesticides at levels deemed "safe" for adults, but exceed safe levels for children, and that pesticide use has increased by ten percent. Because children are smaller and their immune systems less developed, they are more susceptible to and are the greatest victims of pesticide use on food. Carcinogenic pesticide levels have almost doubled since 1993.

Agriculture as we know it today consumes about 17% of all commercial energy used in the United States each year. We need to understand that we must move away from the use of agricultural chemicals for both plant growth and weed control because they consume our energy supplies and quickly deplete the soil of living organisms at the same time. People who promote chemical agriculture either do not understand the complex biological relationships that occur in healthy soil or they have other priorities. Agricultural chemicals promote rapid growth and do not replace what a plant takes away from the soil. They leave behind soil which is devoid of the organisms and/or their residues that give it life.

We must learn, and then teach our children, to use agricultural techniques which return rich, life-sustaining organic nutrients to the soil. Living soil is the essence of agriculture.

There is a lot of controversy surrounding the issue of organic *vs.* nonorganic produce. Many people believe and swear by "organic," saying that produce grown organically has better flavor and greater nutrient levels. In December 1997, the U.S. Department of Agriculture proposed standards for how the word "organic" could be used. Included as accepted organic methods were the use of irradiation, genetic engineering, and sewage sludge. A USDA staffer was quoted as saying, "Officials were 'awestruck' by the protest . . . 200,000 people wrote, faxed, emailed, or spoke up at public hearings . . . it never abated, it just grew . . . We underestimated the strength of commitment to the word organic."

Science argues that most of the completed studies that compare organic and conventional gardening techniques have been too small to address and control all the variables which affect the nutrient levels found in food. John Dromgoole reprinted results of research conducted by Firman E. Bear at Rutgers University in the *Natural Gardeners' Catalog* (1995), as shown in Figure 1. His results seem to disagree with the study-size argument, by asserting that vegetables grown organically have a higher mineral content than conventionally-grown vegetables. Plants obtain their minerals from the exchange complex of the soil. The exchange capacity is usually given in milliequivalents and normally lies between two and 20 milliequivalents per 100 grams of soil.

Figure 1. *Mineral Content of Organic vs. Conventional Vegetables (Firman Bear Report excerpt)*

	Minerals (in milliequivalents)						
Vegetables Type of Soil Management	**Calcium**	**Magnesium**	**Potassium**	**Sodium**	**Manganese**	**Iron**	**Copper**
Snap Beans							
Organic	40.5	60.0	99.7	8.6	60.0	227.0	69.0
Conventional	15.5	14.8	29.1	0.0	2.0	10.0	3.0
Cabbage							
Organic	60.0	43.6	148.3	20.4	13.0	94.0	48.0
Conventional	17.5	15.6	53.7	0.8	2.0	20.0	0.4
Lettuce							
Organic	71.0	49.3	176.5	12.2	169.0	516.0	60.0
Conventional	16.0	13.1	53.7	0.0	1.0	1.0	3.0
Tomatoes							
Organic	23.0	59.2	148.3	6.5	68.0	1938.0	53.0
Conventional	4.5	4.5	58.6	0.0	1.0	1.0	0.0
Spinach							
Organic	96.0	293.9	257.0	69.5	117.0	1584.0	0.0
Conventional	47.5	46.9	84.0	0.8	1.0	19.0	0.5

If the nutritional quality of plant produce does vary considerably from soil to soil, one would expect this to be reflected in the well-being of the people who consume the products from several soils. Not only is the organic production of food good for the environment, it is also nutritionally good for our bodies.

Teaching Laytonville's children to return organic nutrients to the soil began with our experiment in the fall of 1992 to find out if Mary Appelhof's directions for home vermicomposting could be applied to a larger system. These early experiments gave us hope that mid-scale vermicomposting was indeed possible.

Earthworms did help the elementary and middle schools in Laytonville reduce use of both pesticides and natural resources. Avoiding use of pesticides on school grounds and in school gardens is possible because nutrient-rich worm castings applied to the soil maintains soil health.

Figure 2. *Children are the greatest victims of pesticides on food.*

Parents, community members, teachers and students worked together to avoid the use of pesticides on our school grounds. We were concerned about the long term effects of pesticide use on our students. Some teachers felt there was a connection between some of the systemic pesticides being used on our grounds and an increase in headaches and nosebleeds experienced by some of our students. Using vermicompost as a soil amendment helps improve the tilth, nitrogen, and pH of the soil. The improved tilth causes less leaching of nutrients out of the soil and contributes to a healthier soil full of microorganisms, including earthworms, that help increase the yield of crops grown in our school garden. Soil and crops which are healthy are less likely to be attacked by insects and disease. We have also reduced our use of commercial fertilizer as well.

Because the schools relied on worms on-site instead of gasoline-powered trucks for transportation to a recycling plant 25 miles away to process paper wastes, they actually reduced fossil fuel use. Water conservation was another objective of the comprehensive organic gardening program. Mechanical garbage disposal in a kitchen sink requires eight gallons (30 liters) of water to dilute one pound (<0.5kg) of food waste. Because Laytonville schools no longer wash food waste down the drain and, in addition, reuse rinse dishwater to swish out empty milk cartons before transporting them to the recycling center, they used 103,680 fewer gallons (394,000 liters) of water a year after beginning the program than they consumed the previous year.

Above all, vermicomposting reduces solid waste at its source. Every year people in the state of California throw away 146 million tons (132,700,000 t) of trash. Each person is responsible for generating about one ton of waste each year. Of that waste, 880 pounds (400kg) is paper; 200 pounds (91kg), glass; 180 pounds (82kg), metal; 140 pounds (64kg), wood; another 140 pounds (64kg) consists of cloth, rubber, plastic, and leather; and 460 pounds (209kg), yard waste. Schools top the list of sources for discarded paper and food waste. Mid-scale vermicomposting provides a simple, effective, and inexpensive method for processing paper and food wastes that requires no transportation to a central location for further processing. At the same time, mid-scale vermicomposting empowers students who feel they can, and do, make a difference to the positive economies of their town through active stewardship of the natural resources they use.

During the program's first school year, a ten-month period (not a 12-month calendar year), Laytonville vermicomposted 3600 pounds (1636kg) of cafeteria food waste and fed 9360 pounds (4255kg) of protein food waste to chickens and pigs. The schools recycled 567 pounds (258kg) of milk cartons and 654 pounds (297kg) of tin cans. Adding these numbers shows that more than seven tons of solid waste were effectively diverted from one local landfill.

That first year of mid-scale vermicomposting, the school district saved $6000 in dumpster fees by reducing the amount of paper and food waste collected in commercial dumpsters that eventually went to a local landfill. Janitors also spent less time handling trash-can waste. Five years later, this successful vermicomposting program for a school population of 400 students continues to provide a model for student citizenship.

Besides helping reduce the flow of food waste to landfills, mid-scale vermicomposting provides schools with unique study opportunities. An introduction to small business practices arises as staff and students conduct worm workshops and sell both worms and vermicompost to the workshops and sell both worms and vermicompost to the community. Students also experience mathematical concepts and scientific methods close-hand: measurement, ratio and proportion, area, temperature, observing, predicting, testing, comparing, and contrasting. There is no better vehicle than a worm bin for the study of a healthy ecosystem.

A vermicomposting bin is an alive and extremely complicated system. Interdependence, flexibility, diversity, cooperation, and sustainability are all represented in a vermicomposting bin. The inhabitants are so interrelated that to study the system in separate parts is impossible. Worm bins are not just collections of plants, animals, and bacteria. Worm bins are communities involved in complex relationships. Within the walls of a worm bin, millions of creatures are dependent on one another for survival, while others are content to share a desired environment or food source. So, to understand the dynamics of worm bins, we must first look to and understand relationships.

Our combined vermicomposting/organic-gardening program is successful because it is systemic. In life, we can find many types of systems — health systems, educational systems, transportation systems, and so on. Each system is dependent on its many parts which must be organized and integrated if the system is to survive and flourish. Systems thinking, or systems engineering, focuses more on coordina-

Figure 3. *Lunchroom Sorting Scheme*

tion of the total system rather than on its individual parts. Systemic programs are methodical, ordered, and comprehensive.

Vermicomposting is cyclical in nature. The school, a system itself, is in many ways very much like a ecosystem. Energy and matter move through the school community and, like a healthy ecosystem, many substances are recycled. The understanding of connectedness in this process is invaluable, realizing that we do not make decisions in isolation and that our every action has an equal or greater reaction. Actions of each person directly involved in the cycle guarantee the success of our vermicomposting project which does not depend on special circumstances. It is very simply managed by children with the aid of adults. Vermicomposting has become a part of daily life in our school cycle (*see* Figure 3).

Implementing a successful mid-scale vermicomposting program at your site begins with a shifting of old paradigms, and with the vision and willingness to make a difference in your community and in the world. Can your site be a demonstration site? How will you determine what your vermicomposting needs are? Who will be responsible? Who will do what needs doing, and where can you do it? What special equipment and technologies will you need? Where do you look for grant funds? How do you keep records? What happens to the products produced, and who benefits from them? What are the obstacles, and how do you deal with them?

Answers to all these particulars are detailed in the following chapters. This manual aims to be practical guide to help you eliminate the obstacles skeptics may perceive. It uses reading, 'riting, and 'rithmetic to support reducing, reusing, and recycling of natural resources and waste products found in schools and businesses everywhere. Always involve as many people in the initial process as you can. Involvement by others will achieve more "buy-in" and a greater long-term success rate for your program.

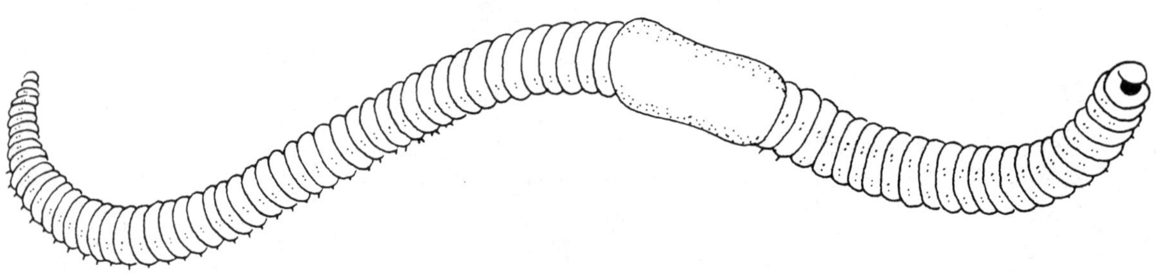

Chapter 2

Determine Your Vermicomposting Needs: Conduct a School-Wide Waste Audit

What is the scope of the problem that you want to fix? Before beginning your vermicomposting project, someone needs to conduct a food-waste audit. The audit will help determine the volume of food waste discarded each day. In school projects, all staff and students need to be directly involved from the very beginning. I suggest the audit be conducted by students with the aid of parents, teachers, or staff. Ideally, students will manage the entire vermicomposting project on a daily basis, as they do quite successfully at my school.

The Cafeteria Staff

Kitchen staff need to be included in the earliest planning phase because they ultimately make the difference between the success and failure of a mid-scale vermicomposting program. Each school site will be different. The cafeteria at Laytonville is centrally located and serves a combined student body of all three elementary, middle, and high schools. Cafeteria food staff were involved in the initial planning stages of the program. Director of Food Services, Pam Cornell, and I met with the kitchen staff and explained to them what my class wanted to try. They gave us their point of view and a list of things which were important to them for the project to be successful. With Pam's input we were able to develop a program which worked for her and the staff. Their ideas and input were crucial to the success of the program and made our project the model it is today.

Then, meetings were held between teachers, janitors, and students. We proceeded slowly and cautiously as to not upset the apple cart. Changes are never easy, but things work when you are all working together towards an end you all share. The kitchen staff were also given much credit for their attempt to reduce the amount of wasteflow being put into our dumpsters. Bottom line, "Share the glory." Talk to one another; good communication is essential for success.

Put two containers that serve as compost buckets for preconsumer food waste in the cafeteria kitchen. As the cooks prepare the food, they will discard their kitchen scraps into the compost buckets. Each container, regardless of location, must be labeled very clearly with words and picture clues.

Clear labeling will insure that plant waste will not be mixed with meat, milk, and cheese wastes. Young children who cannot yet read need picture clues on the containers to know which container to use. Whoever peeks into the food preparation area should also see adults, who are real-life role models whether or not they see themselves that way, using the proper container as well. Your support staff is a valuable resource to reinforce what children learn, and their continued input on program management is very important.

Figure 4.
A Picture-Clue Bulletin Board

At the start of each school year, our kitchen staff spends the first two weeks reviewing the cafeteria recycling procedures with the students. A bulletin board with picture clues is placed in the cafeteria so that each clue shows what kind of food waste should go to which container. The picture clues among Curriculum and Teacher Resources in Appendix G can be used to label your containers and to make a bulletin board.

Initially our food service director stationed her chair near the waste containers to make sure our students put their food wastes into the correct container, while she also collected lunch money. If staff cannot monitor the food-waste recycling, enlist older students or parent volunteers before or after the audit. Today this is only done for the first few weeks of school or until the students are back in the swing of things. Because our students begin composting food waste in preschool, they are well trained and sometimes just need a gentle reminder at the beginning of each new school year.

In my experience, the full success of the project depends on how well it is integrated into the daily activities of the students at school. Often, programs which depend solely on volunteer support cannot sustain themselves. The program needs to become systemic if it is to be effective.

Figure 5. *Labeled Containers*

The Audit

Set up two food waste containers near the trash and recycling containers already in the cafeteria. These containers will hold post-consumer food wastes from students who have already eaten and do their own clean up.

One post-consumer container is for vegetable, grain, and fruit wastes. Earthworms consume a variety of non-meat food waste. This includes all vegetable peelings, fruit rinds, egg shells, breads, pizza crust, coffee grounds, pancakes, oranges, noodles, beans, melons, and much more. The second post-consumer container will hold meat waste and bones, plus dairy products which include cheese and milk.

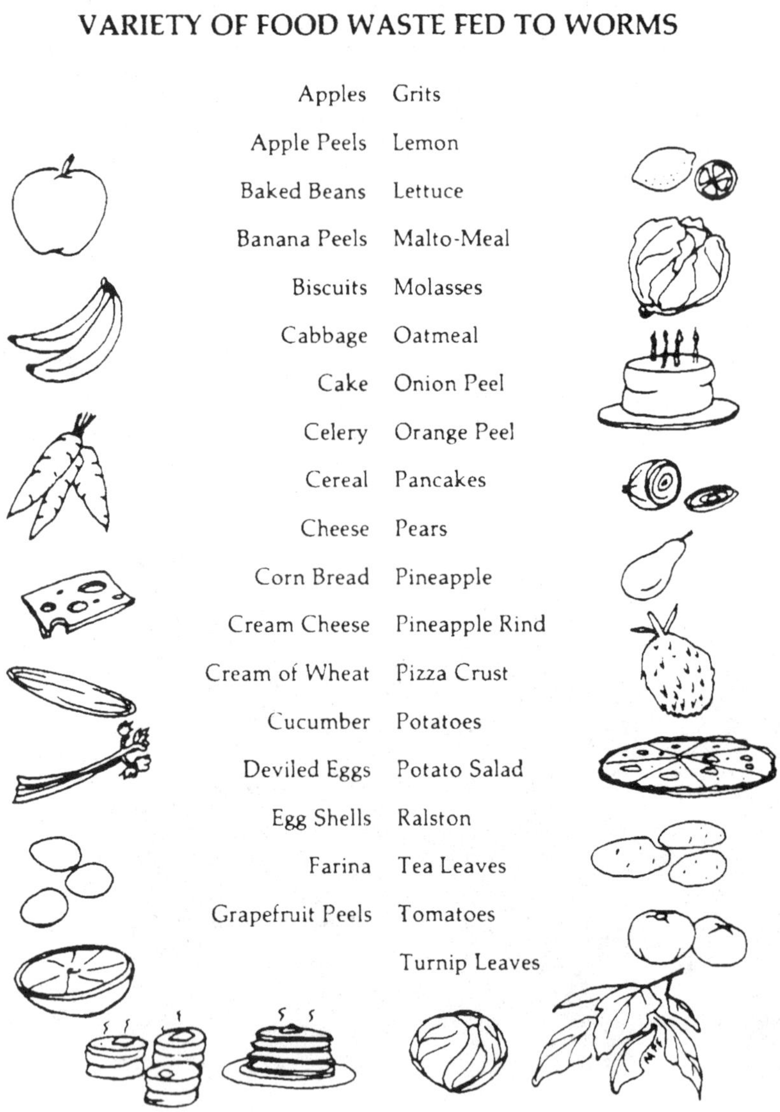

Figure 6.
Chart Courtesy of Mary Appelhof

Earthworms naturally eat both vegetable and animal foods. They can easily eat meat and clean off bones quite well, providing nitrogen for the variety of microorganisms living in the same vicinity. However, *do not* include meat, cheese, or milk in the wastes offered to them at school for several reasons.

We are all familiar with the stench of decaying meat. The odor is caused by putrefaction, the breakdown of protein. Having that type of odor outside of a classroom door is deterrent enough. Coupled with the smell, the process of putrefaction will also attract flies, maggots, and in rural areas, skunks, rats, bears, and other creatures that go bump in the night. Student behavior, however, is the most important reason I have found for not including meat in our worm bins at school. Students do not mind handling the vermicomposted vegetable food wastes, but they absolutely will not touch decaying meat. Occasionally some meat, cheese, or milk will find its way into the bins but it is never enough to cause any problem.

Figure 7.
Weighing Food Waste at School

Each day as the students empty their plates in the cafeteria, they put their food wastes into appropriately-labeled containers. After lunch, students weigh and record the weight of only the food waste (*see* Figure 7) collected from both the pre-consumer kitchen and post-consumer cafeteria locations. In just one month, enough information will be obtained from this audit to calculate the initial system design. On average how many pounds of food waste were collected each day? Add all daily amounts and divide by the total number of collection days. Knowing the amount of worm food to recycle at your site comes before figuring how many pounds of redworms you will need to vermicompost what is generated.

Using Mary Appelhof's simple worm:garbage ratio of 2:1, multiply by two the average weight of food wastes collected per day during the audit to determine how many pounds of worms must be purchased at the start of the project (*see* Chapter 8). Calculating how many bins to construct for their housing can also wait.

After the Audit

Now that you've used audit data to complete the first calculations, double the number of project containers in the lunchroom. Four containers for lunch-generated waste will be placed in the cafeteria-exit path, two each for food waste and nonfood waste. Label all containers with words and picture clues to reinforce their desired contents:

1. Vegetable wastes
2. Protein wastes
3. Empty paper bags
4. Yucky trash

"Yucky" trash includes non-recyclables such as plastic food bags, napkins, and tissues. Health and safety reasons make avoiding the recycling of napkins, tissues, or used paper towels a good idea.

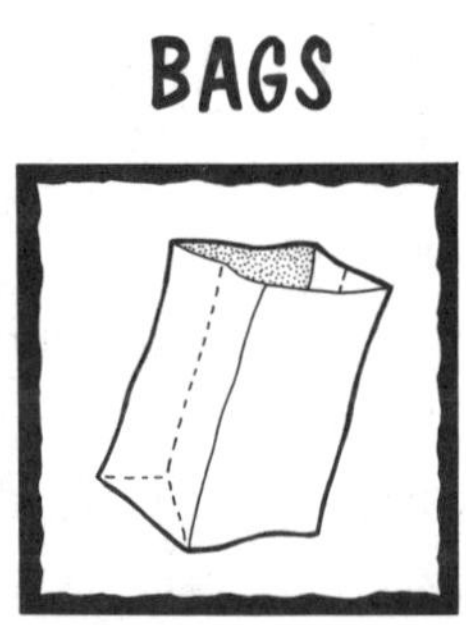

Figure 8.
Containers Needed in the Cafeteria

Recycling containers for glass, paper, and plastic should also be available for non-lunch generated waste. Recycling of these items is a separate project from the recycling of food waste for vermicomposting.

Chapter 3

Bin Dimensions, Materials, and Costs

On Earth every variety of plant or animal is surrounded by a specific environment which is directly related to its survival. As human beings, our homes provide us with an indoor environment which is safe from the harsh physical elements one finds in woods, deserts, countrysides, towns, and cities. Homes reflect the environmental needs of the people who live in them. Housing for worms also will vary, being both artificial and natural. Cardboard boxes with reusable plastic liners, wooden bins, plastic tubs, fishing bait containers, or piles of decaying leaf mold and manure — all can house worms safely.

Many designs and building materials are used to build worm bins. Be sure the bins are not too deep. The design I use is simple, inexpensive, and easy to construct. Our outdoor worm bins were constructed by students, under the supervision of an adult.

When constructing your bin, remember to use treated materials cautiously. Wood that has been pressure-treated with preservatives, for example, will be toxic to the worms and other organisms found in the bin. When wet, preservatives can also leach into the vermicompost. If you use the vermicompost in your vegetable garden, the preservatives could find their way onto your table. On the other hand, salad oil has been used successfully to treat the surface and extend the life of wood used for vermicomposting bins. Salad oil is readily available through many school lunch programs.

How big a worm bin is depends on how much food waste your site generates and how many worms you have purchased to consume it. The information learned from your audit is useful here. Typically, you'll need one square foot (900sq cm) of surface area to bury one pound (<0.5kg) of food waste each day to feed two pounds (1kg) of worms.

Each of our worm bins is a rectangular box (*see* Figures 9-12) approximately four feet wide, eight feet long, and one foot deep (120cm x 240cm x 30cm). We purchased three redwood boards that were one foot wide, eight feet long, and two inches thick (30cm x 240cm x 5cm) for the sides of each worm bin, cutting one board in half for the ends of the bin. Our field studies found that the worms did not mind the redwood which often naturally does not appeal to insects. Laytonville happens to be in "Redwood Country," but redwood is not available everywhere. Substitute cedar or any rot-resistant wood to construct the sides of each bin. Ordinary plywood is all right too. In addition, we purchased two pieces of four-foot by eight-foot (120cm x 240cm) exterior-grade plywood at least five-eighths of an inch (<2cm) thick for the top and bottom of each bin. We didn't buy marine-grade plywood because it was too expensive for school use.

Figure 9. *Worm Bin on Railroad Ties*

If plywood is your choice, make sure it is five-eighths of an inch thick or thicker (at least <2cm). Thinner plywood is not sturdy and is difficult to handle when raising and lowering the lid on the bin. We painted the plywood lid to shed water, but hinging it is optional. Hinged boxes allow the students to handle the lids safely without these covers slipping dangerously onto the ground. The corners of the bins are held together with four equally-spaced 16D nails and screw-in steel-corner reinforcements. The reinforced corners strengthen the integrity of the bins and help keep them from warping. Our bins are in their fifth year of use and show no sign of needing replacement.

Students drilled half-inch (1cm) holes in the bottom of each bin: one hole every nine inches (24cm) across the width of the bin and 18 inches (45cm) apart along the length. These holes allow excess moisture to drain away and also aid in the circulation of oxygen-containing air throughout the bin.

Figure 10.
Bin Holes

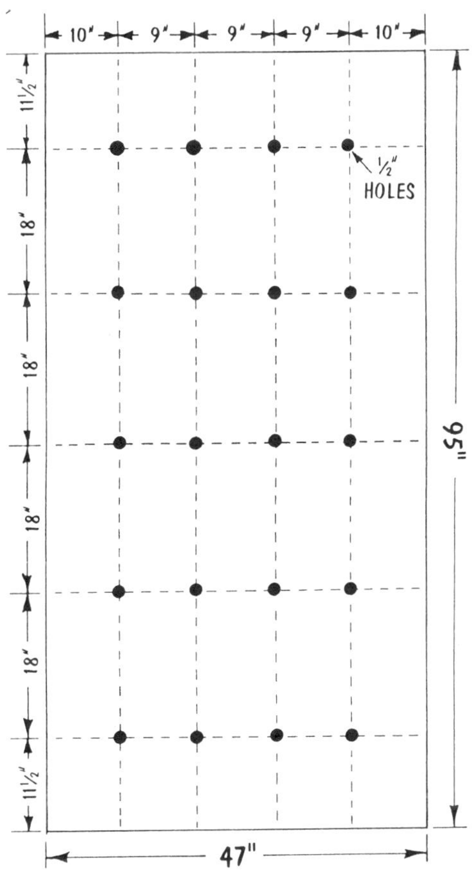

Placing the bin on used railroad ties keeps it off the ground for better drainage and for good air circulation inside the bin when it is filled with worms and bedding. One railroad tie is placed under each long side of the bin, and one tie supports the middle of the bin. Cinder blocks or bricks could substitute for the same purpose.

My husband worked with six capable middle-school students who constructed the bins, drilled holes in the bottom of each, and attached the steel-corner reinforcements. The students practiced a little at driving nails beforehand and had good direction. It took them about two hours to complete one mid-scale worm bin that cost $100 in materials to construct for school use.

We used student council funds to purchase the materials for all four mid-scale worm bins. They cost Laytonville Elementary and Middle Schools $400 total. In ten months' time we eliminated the use of one trash dumpster to save $6000 in disposal costs. Our bins have more than paid for themselves and have reduced transportation and disposal costs for paper and food waste. Subtracting what we paid for materials, our net savings was $5600.

Our outdoor worm bins also incorporate four provisions for successful mid-scale vermicomposting:

1. Introduction of oxygen
2. Regular turning
3. No compaction
4. Students as both labor and management

Successful mid-scale vermicomposting is dependent on a good oxygen supply, not only for the worms, but also for the aerobic microorganisms and other decomposers at work in the bin. A worm bin in which the layers of bedding and food waste become compacted develop foul odors quickly, due to the gas produced by the presence of anaerobic bacteria. These anaerobes will slow the decomposition process. Remember, a properly maintained vermicomposting system does not smell bad. On the contrary, vermicompost has an earthy odor similar to that of composting vegetation on the forest floor. All of our mid-scale worm bins are located outside, near my classroom. Complaints about bad smells did not surface. One would never know that in ten months' time, worms had successfully composted 3600 pounds (1620kg) of

Figure 11.
Turning Fluffy Bedding

cafeteria food waste without objectionable odors just outside the classroom door.

Compaction of food waste may occur in a worm bin whenever turning the bedding becomes difficult because either of these conditions occurs:

1. A bin is too deep.
    a. The combined weight of food waste and bedding favors compression.
    b. Students struggle with bringing the bedding and food waste from the bottom of the bin to the top of the bin.
2. Bedding is in large pieces that are not fluffy; it packs down quickly.

Shallow bin depth of no more than 12 inches (30.5cm) and frequent turning one or more times weekly are the best ways to prevent compaction. Shallow bins help prevent the

Figure 12.
Shallow Bin Construction

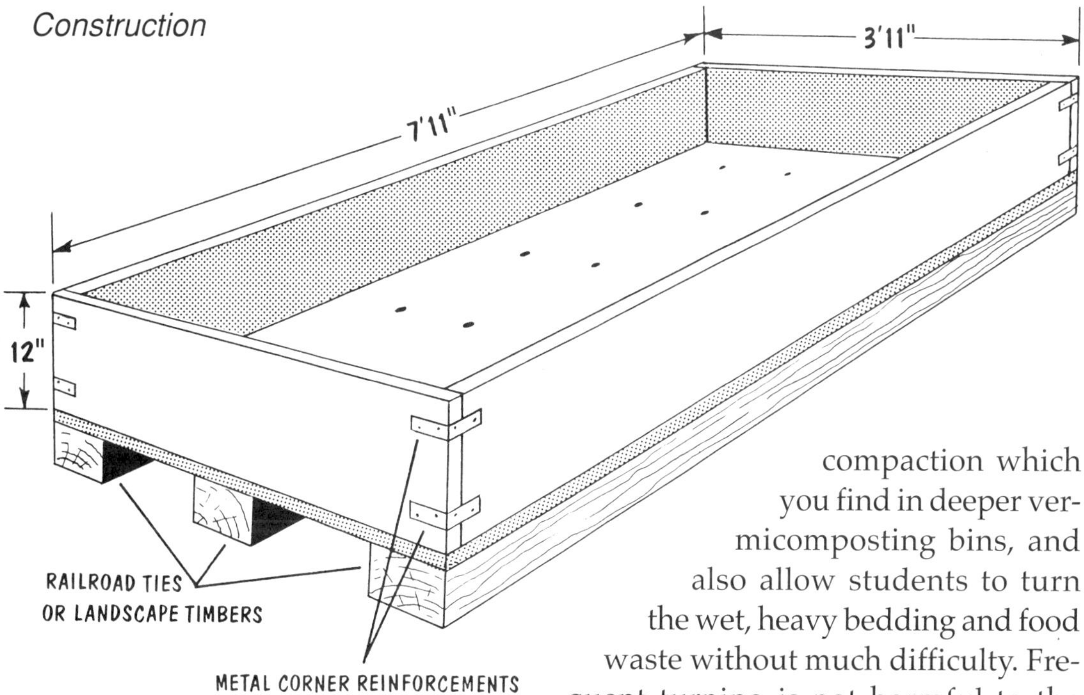

compaction which you find in deeper vermicomposting bins, and also allow students to turn the wet, heavy bedding and food waste without much difficulty. Frequent turning is not harmful to the worms and introduces more oxygen into the bedding. Oxygen is essential for aerobic bacteria, other decomposers, and worms to work at their full composting potential. However, natural compaction of vermicompost will occur even in shallow bins, as the worms convert more and more of it into castings.

Mid-scale vermicomposting bins are similar to those for home composting with a few exceptions. In small home systems the food waste is buried and left to sit for two to six months, while the contents are turned more frequently in mid-scale systems. Home systems can be kept indoors and are more sensitive to outside temperatures because they have less mass to insulate them against fluctuating temperatures. Mid-scale systems are kept outdoors and are more forgiving of extreme temperatures because of their mass.

Appelhof's 1′ x 2′ x 3' (30x60x90cm) home vermicomposting worm bin will compost about a half to one pound (<0.3-0.5kg) of food waste a day. Our mid-scale worm bin will compost approximately 12 to 24 pounds (5.5-11kg) of food waste per day. The efficiency of a worm bin to convert food waste to castings is determined by air, environment, temperature, moisture, surface area, and health of the worm population.

Chapter 4

Location Location Location

Where can vermicomposting take place? What environments are best? Our worms do not seem to mind nearby pounding and reverberation from footfall, frequent turning, or exposure to students everyday; in fact, they seem to thrive. Worms are photosensitive and do not like light, although some feeding experiments have desensitized them to light. Consider that worms cannot tolerate temperature extremes, neither high or low. Temperature affects both worm survival and feeding patterns on food waste, and is the greatest environmental factor in how fast food waste is converted to vermicompost. Thermometers both inside and outside our bins provide useful temperature information.

Figure 13.
Thermometer Inside Bin

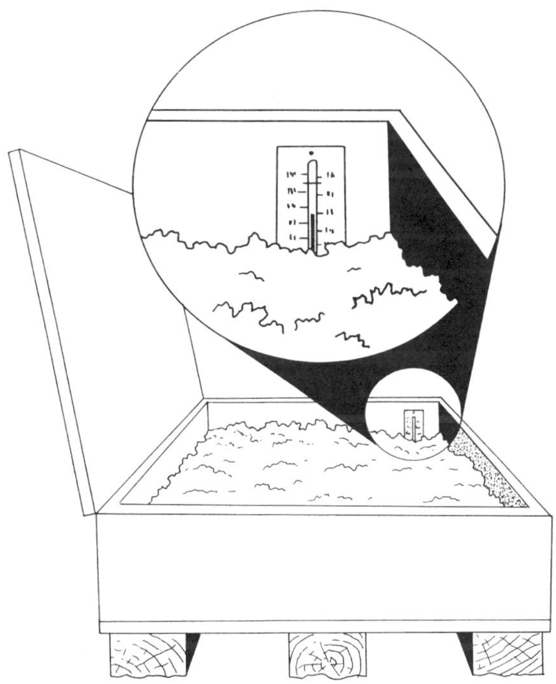

When looking for a place to locate your worm bins, remember that the bins need protection from extremely high temperatures. Temperatures above 90°F (32°C) can be harmful to earthworms. Protection from high temperatures can be accomplished by placing the bins in a shaded area or by covering them with lath or shade cloth. Our worm bins are located next to a building and are in the shade for the greater part of the day. Being able to prop open the lids against a building (*see* Figure 14) is helpful for students who sometimes have difficulty with the weight of the plywood lids.

Mid-scale vermicomposting systems, such as our 32 cubic-foot (1 cubic-meter) bins, are more tolerant of extreme outdoor temperatures than are smaller two to four cubic-foot (0.05-0.10 cubic-meter) bins that we use in our classrooms. Larger bins have a greater insulation capacity because they hold a larger mass of paper bedding. In the summer, earthworms can be kept in a shaded location outside as long as the air temperatures aren't consistently over 90˚F (32˚C). We use a large amount of damp bedding, created from discarded paper and added water, in our bins that keeps the temperature in the bedding considerably cooler than the summer air temperature outside of the bins. The temperature is cooler because the process of evaporating water removes heat from the bin in the same way that wearing wet garments cools us in the summer.

Figure 14.
Building as Support for Open Lid

One summer our worm bins were moved by forklift away from the shade of the building and into full sun for building maintenance reasons. However, to protect them from both the sun's heat and light, we installed shade cloth, a lightweight mesh fabric purchased at a local nursery (*see* Figures 15 and 16). Much to my surprise, the shade cloth was very effective in protecting the worms from daytime highs of 90˚-100˚F (32˚-38˚C).

We observed that feeding patterns were slowed, but not adversely affected, by dropping outside temperatures as long as the bins were full of paper bedding and receiving food waste. Slowness in feeding can cause food waste to accumulate in the bin. Adding additional bins can pick up the slack caused by the slower rate in feeding.

Figure 15.
Protective Shade for Worms

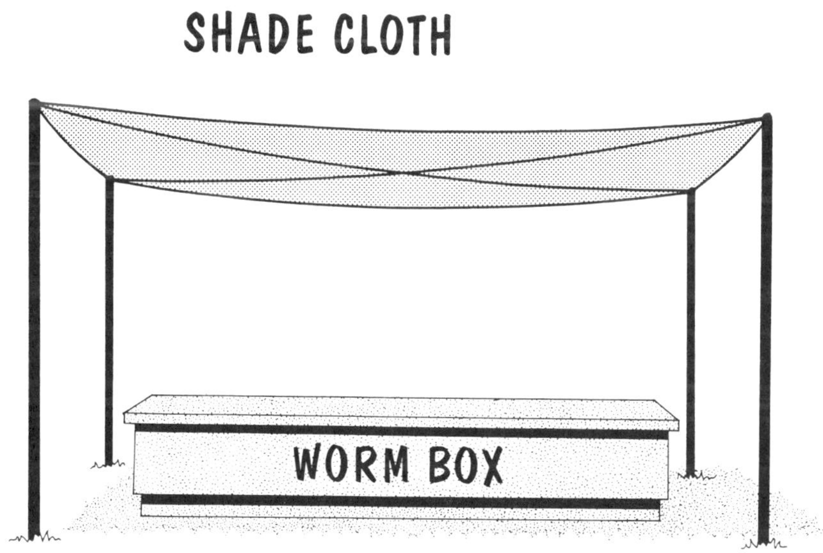

According to the laboratory experiments conducted by Dr. Roy Hartenstein, optimum feeding occurs at temperatures between 68°-77°F (20°-25°C). Laytonville's elevation is 1200 feet (360m) above sea level. Temperatures often fluctuate from cool

Figure 16.
Shade Cloth Support

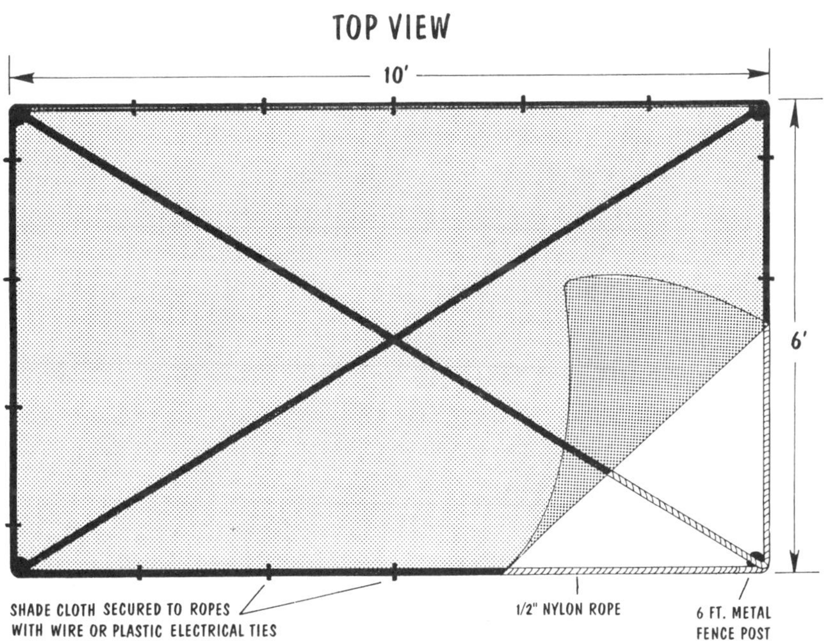

50°F (10°C) mountain nights to hot 100°F (38°C) days. We rarely experience temperatures below 10°F (-12°C) in the winter or above 105°F (40°C) in the summer.

Even with our fluctuating mountain temperatures, the worms continue to feed throughout the winter months. At times, the surface of the paper bedding may become slightly frozen. Because the worms are feeding on buried food waste, this is not a problem. Only in the coldest part of winter is feeding slowed in certain areas of the bin, such as around the edges of the bin where the worms are less insulated. You can insulate the outside of the worm bins with hay bales or any type of standard insulation material to help maintain warmer temperatures and constant feeding.

When nighttime temperatures drop into the teens, and continue to stay so for a month or more, the top quarter inch (0.5cm) or so of paper bedding freezes and the worms move deeper into their bedding. We found that the small amount of naturally-occurring composting of vegetable waste created enough heat to keep the temperature of the bins warmer than the surrounding air temperature. We also observed that the worms would migrate to warmer areas in the bins where microbes were hard at work and food-waste composting was taking place. The larger size of our outdoor mid-scale bins (32 cubic feet, <1 cubic meter) provides much greater insulation than smaller indoor bins (two to six cubic feet, <0.1-0.2 cubic meters). This insulation in winter conserves the heat produced from the composting food waste so that the worms are more active and consume more food than if their bedding is near freezing.

Because of their size (32 cubic foot, <1 cubic meter) and because of the moisture which drains from them, our mid-scale bins are located outside. The excess moisture comes from food waste and sometimes over-wet bedding. A mid-scale bin could be located indoors if you had a system for collecting the accumulated moisture, and then provided an area for the moisture to run off.

Smaller bins are frequently kept indoors. These indoor mini-units are extremely important because they can be used in the classroom during inclement weather for training. During the winter months, when my students cannot get outside to study earthworms, we work indoors in the portable bins and study earthworm science. Using the classroom curriculum guide, *Worms Eat Our Garbage,* my students become worm experts in preparation for the spring when they will teach worm science to the preschool through fifth grades, as well as give tours to guests from visiting schools.

In the spring, the bins and experts together can be moved into other classrooms. Small worm bins can also leave our school site, becoming an integral part of student homework. Students take the mini-bins home for one week to educate their families on the principles and benefits of vermicomposting.

In our experience, the mini-units are too small to keep outside. They do not have enough room for the insulation mass needed to protect the worms from extreme outdoor temperatures winter and summer. The worms become sluggish and feeding slows. When left outside for prolonged periods of wintertime, the bedding freezes and so do the worms. As long as temperatures aren't consistently above 90°F (32°C), they can be kept in a shaded location outside. Location of your bins mostly depends on their size and the air temperature around them.

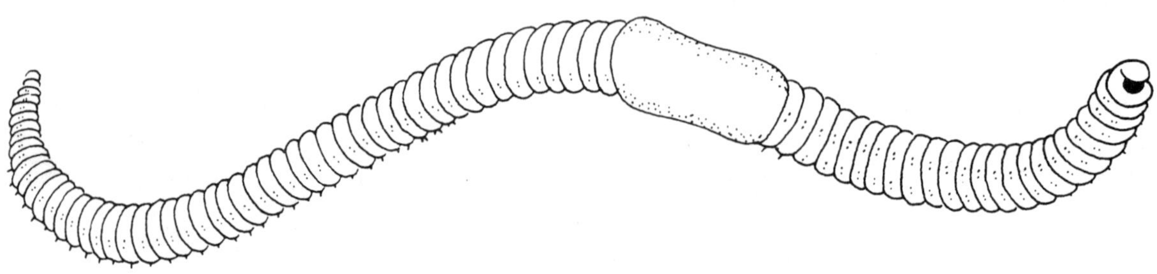

Chapter 5

Beddings: Paper Paper Everywhere!

Bedding is a moisture-retaining medium placed inside the bins and functions as housing around the composting worms. Worm beddings are important because they provide four environmental requirements for earthworm health and safety:

1. *Moisture*. An earthworm needs a moist environment in which to live. Earthworms exchange oxygen and carbon dioxide through their skin and, for this transfer of gases to occur, their bodies must be kept moist with water present in their surroundings.
2. *Oxygen-rich air*. Worms require oxygen from the air. Loose bedding with many air pockets allows respiration to occur and prevents worms from suffocation.
3. *Protection*. Bedding protects worms in the same way that soil protects soil-dwelling earthworms or leaf litter shelters worms in the woods. Bedding covers the worms and provides a medium through which the worms move. It hides worms from predators, such as toads, snakes, or birds, to give them a competitive survival advantage.
4. *Food*. Worms eat the bedding, as well as food waste buried in it, although bedding by comparison has little nutritive value. As an alternative food source to eating their own castings, which can be toxic to them, bedding is a burial medium that gives worms the advantage of a safe food source.

Many beddings can be used: manure, shredded cardboard or paper, straw, wood chips, grass clippings, sawdust, or peat moss. I do not recommend using peat moss because it is a nonrenewable resource. Ideally bedding should be light and fluffy to allow sufficient transfer of oxygen throughout the bin, as well as to avoid compaction.

Paper Wastes

Schools throughout the nation generate more food waste and discarded paper waste than any other type of waste. We deliberately chose to use some of our discarded paper waste for the worm bedding. The paper we now use for bedding consists of discarded newsprint paper, book orders, and colored construction paper from all classrooms in Laytonville's preschool, kindergarten, and grades one through eight. Our higher-quality Xerox paper is sent to the recycling center along with our glossy clay-coated papers. The clays make paper glossy and retard the absorption of water. They can contain potentially toxic metals, such as cadmium, chromium, and lead, which can be harmful to humans as well as animals. Recently, the paper industry has also used these clays in the de-inking process during paper recycling.

Each classroom is responsible for collecting and sorting its own paper waste. Located in each classroom are three recycling boxes to collect different kinds of paper separately:

1. High-quality Xerox paper
2. Glossy paper and magazines
3. Newsprint and colored construction paper

Students, using wheelbarrows, wagons, or carts, pick up the paper from each classroom once a week. The Xerox and glossy papers go into the back of a janitor's truck for delivery to the recycling center; staff or volunteers transport them to the recycling center. I recommend using staff for this task and saving all the worm-related parts of the vermicomposting project for student workers.

It is important to remember that children learn by doing. Students need to be engaged in actions that broaden their own personal experiences. As student wormworkers, they become involved in meaningful experiences which reconnect them with the Earth. We need to view students in terms of what they are capable of doing. For teachers it means doing things differently, teaching and facilitating learning in new ways. We need to reverse the feeling of hopelessness that so many of our students harbor to those of hope and healing. Saving all the worm-related work for the students is important in helping them understand the connection between action, learning, and change.

Other non-school sites send us their discarded paper. Our district office sends us its shredded confidential information for vermicomposting, and our local medical clinic recycles its shredded documents at our school sites. Any business which handles confidential information is a good paper source. We do not use all of our discarded paper for bedding, but we use a lot of it. Roughly two-thirds of our paper waste is handled on our site. Cardboard is sent to the recycling center.

If a school vermicomposting project is being planned, please consider using on-site paper waste as bedding in the worm bins. It only makes sense and reduces waste close to its source.

The Shred Shed

Originally I had students shred the paper in my classroom. I found that the dust created by shredding paper interfered with the workings of my classroom's computers. We needed an area to organize, process, and contain the paper outside of my classroom. Cooperation came from a high school shop class that built us a Shred Shed, an eight by ten-foot building (about 7.5 square meters) perfectly adequate for our needs (*see* Figures 17-21).

Figure 17.
Shred Shed Designed and Built by High School Students

Students deliver the paper collected for worm bedding to our Shred Shed. This is a small building maintained for storing and shredding paper; however, any room can be designated for these purposes. Along each wall inside the Shred Shed are storage containers which we constructed using polyvinylchloride (PVC) pipe (*see* Figures 18-21). Economical at ten cents per foot and easy to assemble, the PVC structure supports all the separate bags which we use for storing the unshredded and shredded newspaper and construction papers until they are needed as bedding.

The fittings to put the PVC pipe together cost about $10. The glue was about one dollar, and the black pipe (used to clip the storage bags onto the frame) also cost one dollar. Materials to make our storage containers cost only $20 total. The

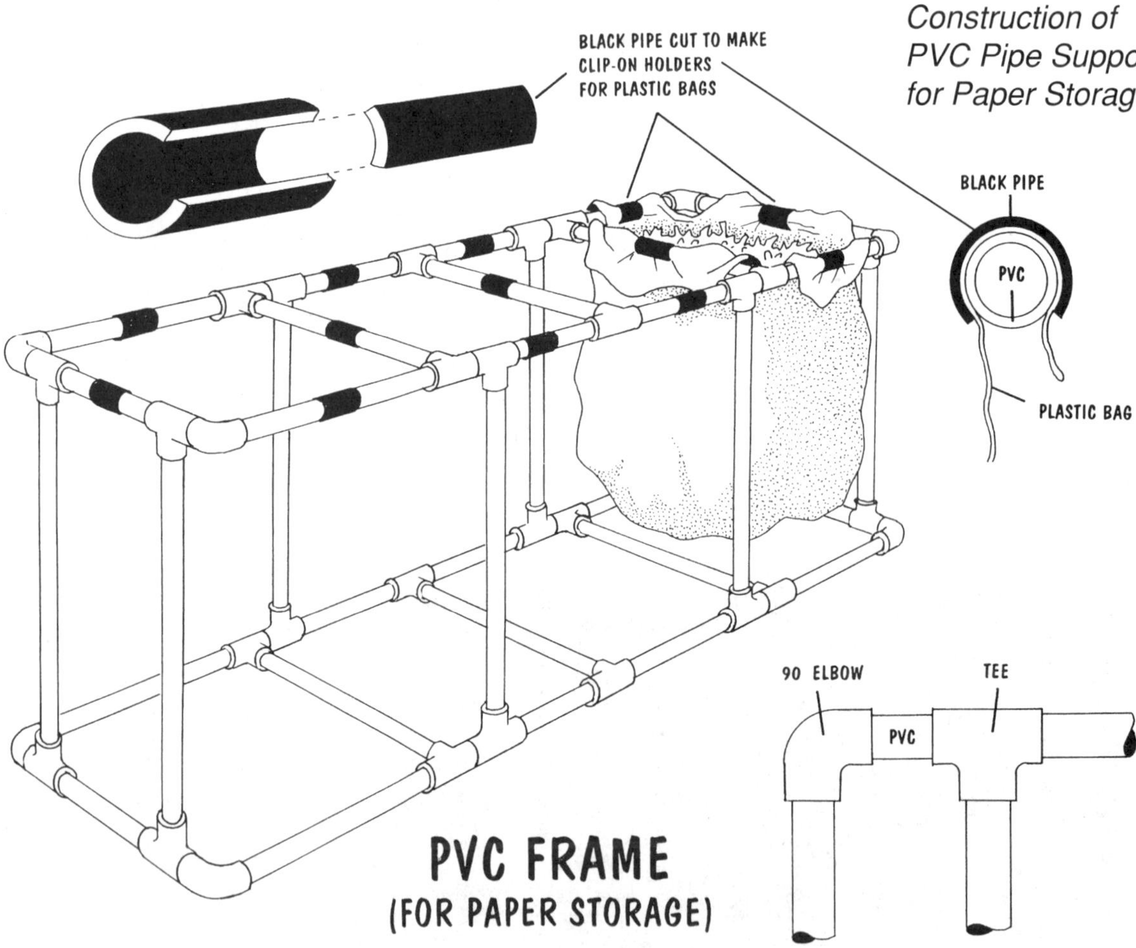

Figure 18.
Construction of PVC Pipe Supports for Paper Storage

storage bags clipped to them can be made from fabric or you can use 50-gallon (188-liter) plastic trash bags. We use plastic trash bags because they are available at school from janitorial supplies. Plastic bags are reuseable, and they shed water in

Figure 19.
Shred Shed Layout

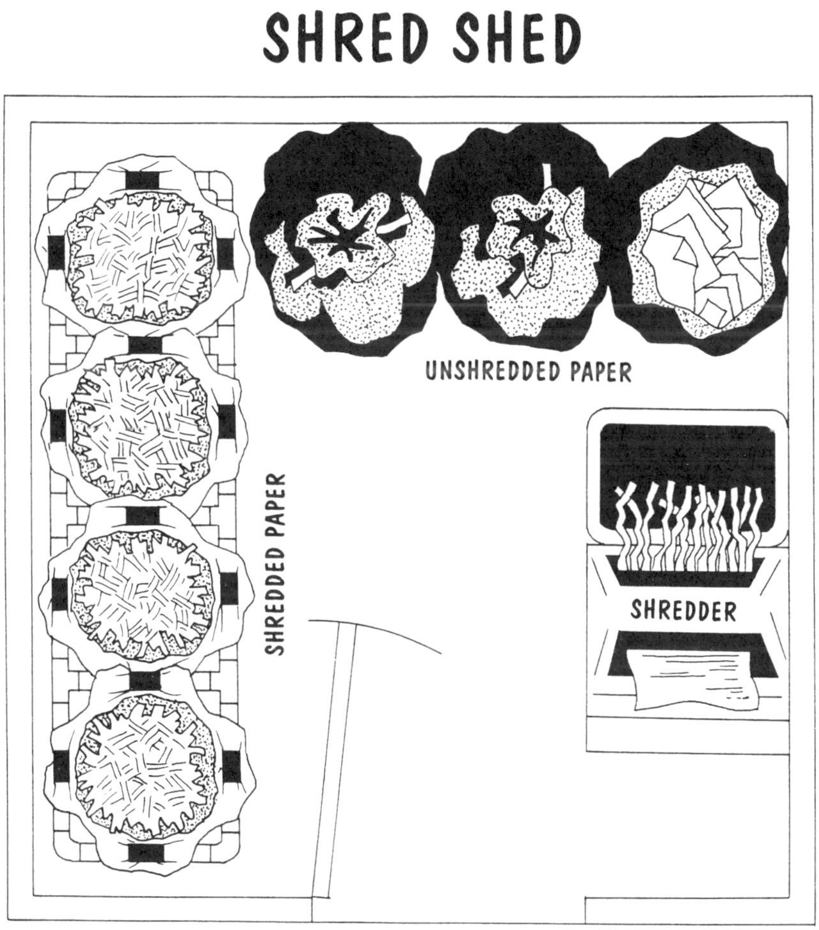

the event the Shred Shed door is accidentally left open to allow wind-blown rain inside. Plastic trash bags can usually be purchased at office supply, grocery, and convenience stores.

While the project was still new, the students tore up the paper by hand. This may be the way to start, but there are real advantages to purchasing a powered, office-type paper shredder. We found out very soon that hand-torn paper settled and compacted easily. The compacted layers of paper impeded the circulation of oxygen throughout the bin, allowing anaerobic bacteria to multiply. This slowed down the process of vermicomposting. Secondly, hand-torn paper often became stuck on the tines of the turning fork, proving difficult and frustrating for the students to handle. It did make some difference as to the kind of turning fork they used. Potato forks with their rigid blunt tines are far superior to hayforks (pitchforks) which have narrow, sharp, pointed tines. Potato forks do not bend, and they easily scrape the bottoms of bins.

Machine-shredded paper is fluffy and therefore easy for students to turn. Its fluffiness also allows for excellent air circulation, an important bedding criterion. Use of machine-shredded paper gives another advantage; its small size promotes efficient biodegradability, turning more quickly into useable vermicompost. Any material which living organisms break down into simpler components is biodegradable. Pure newspaper or newsprint biodegrades faster by itself than when it is mixed with construction paper. It is thinner than construction paper to begin with and is not treated with chemicals that retard water absorption.

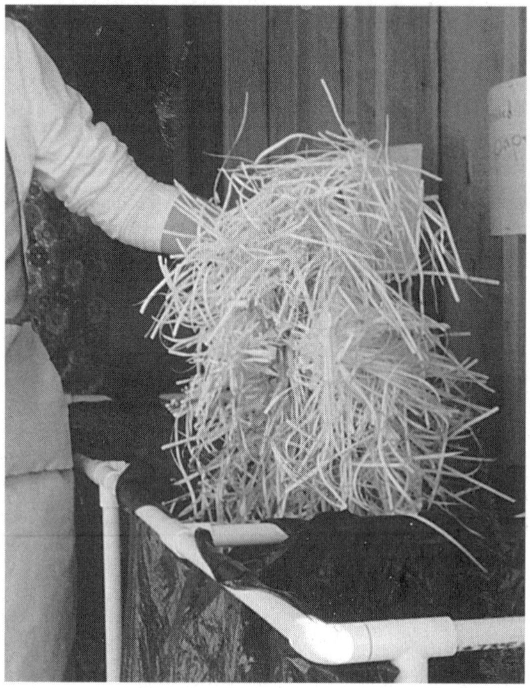

Figure 20. *Machine-Shredded Paper*

If money to purchase a shredder is not immediately available, by all means, have the students hand-shred the paper until the district or a funding authority saves enough money on dumpster fees to afford to buy a good shredder. Everyone will appreciate better its ease of operation and effectiveness after they have tried other methods first.

Paper shredders ideally will be located inside a Shred Shed or its equivalent. Our power company, Pacific Gas and Electric (PG&E), annually gives mini-grants to classroom teachers for projects which reduce consumption and conserve energy. Find out if these grants are available in your area by talking with local power company representatives. I used a PG&E grant (*see* Appendix B) to purchase two small shredders from a large office supply store, for about $100 each, to upgrade the efficiency and effectiveness of our project. Eventually we purchased one larger shredder for about $1000. Using the larger shredder took less time than using both small shredders together.

In our district, students in the fourth through twelfth grades use electric shredders to shred paper. They must take a power tool safety test first and work with a student who has already been trained to use the machine. We have a shredding schedule that I create with staff input each year. It is posted in the school office, classrooms, and staff room. Students often come out to the Shred Shed at recess to shred paper as well, so there is never any lack of bedding for our worm bins. If we ever have an excess of shredded paper, there are always people who pick some up to take home for their own worm bins

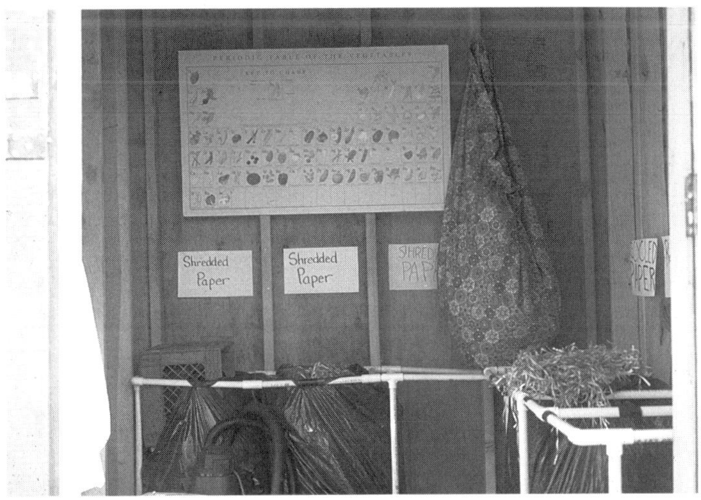

Figure 21. *Inside the Shred Shed*

Students are responsible for monitoring the worm bins and adding more paper bedding as needed. When the vermicompost is turned, additional bedding may need to be added to cover food waste brought to the surface. Eventually the worms will consume the bedding along with the food waste. The paper bedding has very little nutritive value for the worms, but provides both a place for us to bury food waste and an environment for our worms that supports their physical needs.

Chemical Concerns

When I started the project, I was concerned about the toxicity of inks found on the newspaper, since most inks were derived from petroleum solvents or oils prior to 1986. The effect of petroleum derivatives on laboratory animals made their untested effects on people worrisome. Not much is known about accumulated petroleum and dyes in the soil and how they affect plants and animals that ingest them. Some pigments can hold potentially toxic metals. Cadmium, chromium, and lead are sometimes blended in as colored and black ink fixatives. The new soy-based inks are safer.

In 1986 the American Newspaper Publishers Association (ANPA) began to encourage the use of inks made from soybeans instead of petroleum. Currently, almost half of U.S. newspapers print with soy-based inks. A newspaper's customer service or production department may be willing to describe local practice during a phone call. Group visits may also be informative.

Figure 22. *Toxicity Report (excerpt)*

Alpha Analytical Laboratories Inc. • 860 Waugh Lane, H 1, Ukiah, California 95482 (707) 468-0401

CHEMICAL EXAMINATION REPORT

Laytonville Elementary School
P.O. Box 325
Laytonville, CA 95454

Attn: Binet Payne 9/15/93

Batch 93-0901-005 Consisted of 1 sample and 5 tests.

Sample 1 Laytonville Elem. Sch — Worm Castings
Sample Type: Soil Sampled by: Binet Payne
Sampled 8/31/93 Tested 9/11/93

	Method	Result	Units	MDL
Benzene	EPA 8020	ND	ug/g	.005
Toluene	EPA 8020	.0076	ug/g	.005
Ethylbenzene	EPA 8020	ND	ug/g	.005
Xylenes	EPA 8020	.031	ug/g	.005
Lead	7420	ND	mg/kg	5

MDL= Minimum Detection Rate (action level)
ND = None Detected

Toxicity of colored papers themselves was also a concern because dyes used to produce them are often derived from benzene, toluene, naphthalene, and other aromatic hydrocarbons. These chemicals are known carcinogens.

Before I began using the vermicompost produced in our bins directly on our school vegetable garden, I wanted to be certain that it was safe. I had the compost tested by Alpha Analytical Laboratories of Ukiah, about 50 miles (80km) south of Laytonville. Samples were tested for benzene, toluene, ethylbenzene, xylenes, and lead. Alpha Labs found no traces of benzene, ethylbenzene, or lead. They did find toluene and xylene at minimum detection levels which, according to the lab's explanation, were insignificant and therefore not harmful.

Whatever company supplies the colored papers purchased by your school should be able to verify the presence and level of petroleum derivatives in them. Beckley-Cardy is our supplier and provides this information through its toll-free 800 number, 1-800-446-1477.

All paper products are not the same. Be particularly cautious with what you use.

Alternative Beddings

Peat moss, animal manures, straw, leaves, and cardboard are alternatives to paper bedding materials. Other schools in California are currently using wood chips and sawdust as beddings in their worm bins. Grass clippings are also among the possibilities. Regardless of their source, be sure your beddings are non-toxic.

Many times lawns are treated with chemicals which are toxic to worms. One of the most common herbicides used on lawns is 2,4-D which is highly toxic and contains traces of dioxin. If the beddings you use are laced with pesticides or herbicides, the resulting vermicompost can carry these chemicals into your garden. Because little is known about the transfer of toxins through vermicompost to your garden and vegetables, it is better to make sure you are not using hazardous materials in the process. Avoid any materials you can't confirm are safe. Researchers at the University of Kentucky discovered that a single application of common pesticides — benomyl, ethoprop, carbonyl or bendiocarb — at recommended label rates killed 60%-90% of the earthworms in test plots. Effects of these chemicals on people may not be known, but it is better to be safe than sorry. If vermicompost produced in herbicide-laced bedding is used in vegetable gardens, no one can predict accurately the effect such chemicals will have on people who eat produce from those gardens.

A potential disadvantage of using animal manures and green yard waste as worm bedding is that they spontaneously begin to compost. Composting is the natural process of decomposition which involves recycling of materials into a rich humus for use as a soil amendment when applied directly to the garden. In the presence of the proper 30:1 carbon to nitrogen ratio, the composting process will begin. Composting materials heat up to temperatures above 160°F (71°C), and temperatures anywhere above 84°F (29°C) can be harmful to your worms. Temperatures above 86°F (>29°C) are stressful and at 95°F (35°C), they may die. To prevent high temperatures during vermicomposting, use manure, grass clippings, and green leaf matter sparingly. You can monitor the temperatures at different places in the bins using a simple compost thermometer. If you find a bin which is heating up, simply mix the contents with more paper bedding and this should cool things down.

Resources for worm beddings vary from site to site, and there are many beddings which can be used quite successfully. You must decide which worm beddings among those available to you will best serve your individual needs. Our experience in Laytonville has demonstrated that discarded, shredded paper is plentiful, economical, and useful. If you work in a school setting, I strongly recommend using paper to vermicompost your food waste. I also look forward to learning from other schools and businesses that experiment and document their alternatives to paper beddings.

Chapter 6

Earthworms Inside and Out: Anatomy of a Compost Machine

The organism central to the biological technology of vermicomposting is the earthworm. Earthworms are animals in a group we know as invertebrates, which have no backbones. They are in the phylum Annelida and are commonly called annelids. Included in this phylum are also marine worms and leeches. Their name comes from the Latin word *annelus,* which means ring, so it is not surprising that an earthworm's body looks like it is made of many rings. Between each two rings is a segment.

Earthworms are exceptionally valuable to the environment. Without their constant burrowing, the soil would lack good drainage and aeration, and their nutrient-rich castings would not be mixed into its upper layer. Topsoil would not be as rich in nutrients and certainly would differ in consistency. Few places on earth do not contain earthworms, but nutrient recycling occurs much more slowly in these places. These master decomposers turn figurative mountains of rotting plants and animals into nutrient-rich soil. Earthworms are Nature's recyclers. With truly flexible musculature from "head to tail," earthworms are highly adapted for living underground.

Earthworm Species

Of more than 15,000 species of segmented worms that live in our world, over 4,400 are different species of earthworms. A species is defined as a basic biological category of related individuals capable of breeding among themselves. Names for species follow their genus names. For example, while most earthworms are reddish in color, *Allolobophora chlorotica* is green. *Allolobophora* is the genus name, beginning with an uppercase letter, and *chlorotica* (all lowercase letters) is the descriptive species name.

Whenever I refer to earthworms for vermicomposting in this text, please understand that I am *not* writing about nightcrawlers. Nightcrawlers are subsoil dwellers who live deep below the surface and bring minerals found in those depths closer to the soil's surface. They prefer to spend their time alone in burrows, except when searching for a mating companion. Nightcrawlers are commonly found in U.S. gardens and will grow to 10 inches (25cm) in length. Their scientific name is *Lumbricus terrestris*.

One day a student brought a nightcrawler to the classroom and placed it in our mini-vermicomposting bin. Later that day, during a parent conference, I noticed this huge nightcrawler making its way along the floor and out the classroom door. It obviously did not want to make a home with the rest of the earthworms living in our small classroom bin. Escape and a hasty retreat meant survival.

"A worm's gotta do what a worm's gotta do."

The giant Australian earthworm *Megascolides australis* can reach six feet (2m) in length. Its diameter can be one inch (2.5cm) around. South Africa can boast the largest earthworm ever found — a whopping 22 feet (<7m) long, extending across the road where the worm was seen.

The earthworm species used in mid-scale vermicomposting has many names. Manure worm, redworm, or red wiggler are just a few of its common names, but its scientific name is *Eisenia fetida*. Average size for *Eisenia fetida* is 3.5 inches (9cm) long and one-eighth inch (0.3cm) in diameter. These creatures are communal and thrive under cultivation in a confined space, as long as environmental conditions favor their survival. I never have had redworms leave a container located either inside or outside my classroom, unless they were left in the sun or the conditions inside the box were so unhealthy they were forced to escape to survive.

Knowledge of earthworm biology helps students understand why they must treat the earthworms in their care respectfully.

External Features

Although specific details vary for different species, earthworms have many features in common. Skin-like epidermal tissue covers the body of an earthworm, and a long smooth and swollen "saddle" appears nearer its forward end. How-

Figure 23. *External View of Earthworm Anatomy (Lumbricus terrestris)*

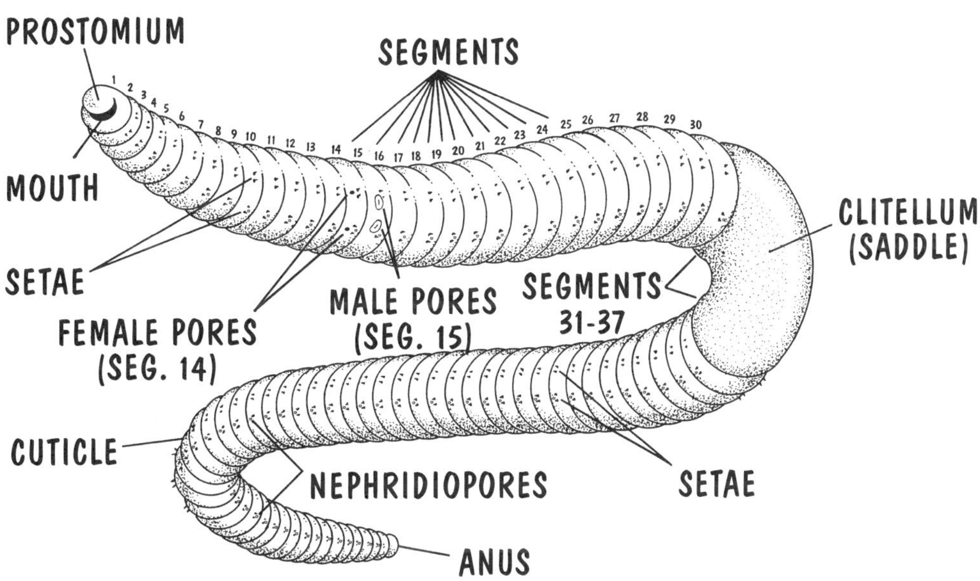

ever, the most notable feature of the earthworm's body between its forward mouth and rear anal opening, is its division into about 150 segments. Each segment, except the first and last, has four pairs of stiff bristles which are called setae (*see* Figures 23-26). The setae are extended or retracted by the earthworm's protractor and retractor muscles to help the worm move forward. As the body stretches forward, setae

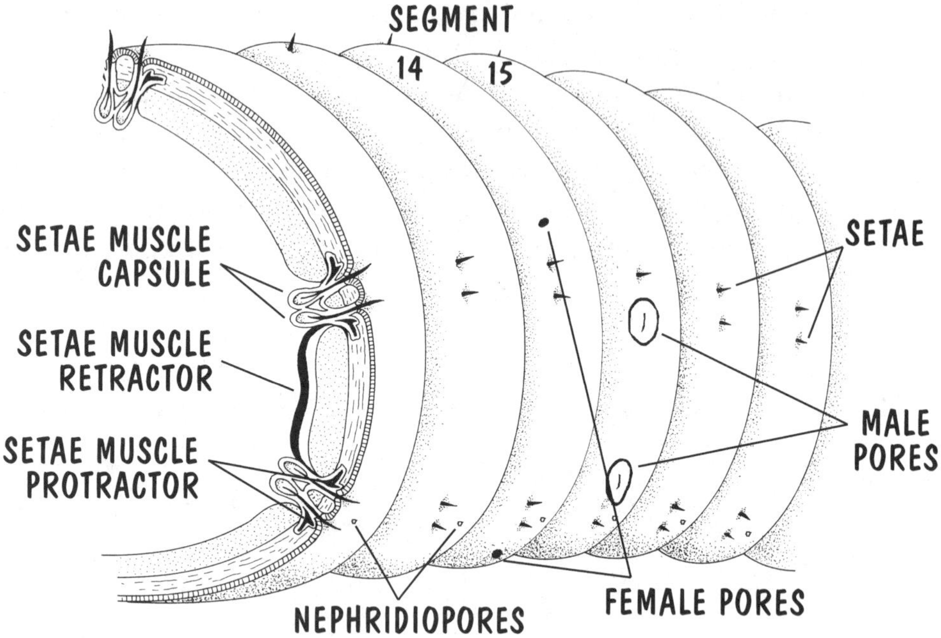

Figure 24. *Ventral Surface (counterclockwise rollover)*

anchor it to the ground and allow the rear part of the body to catch up with the rest of it.

Internal Systems

A cross-section of an earthworm looks like an archery target of three concentric circles. An outer ring begins with

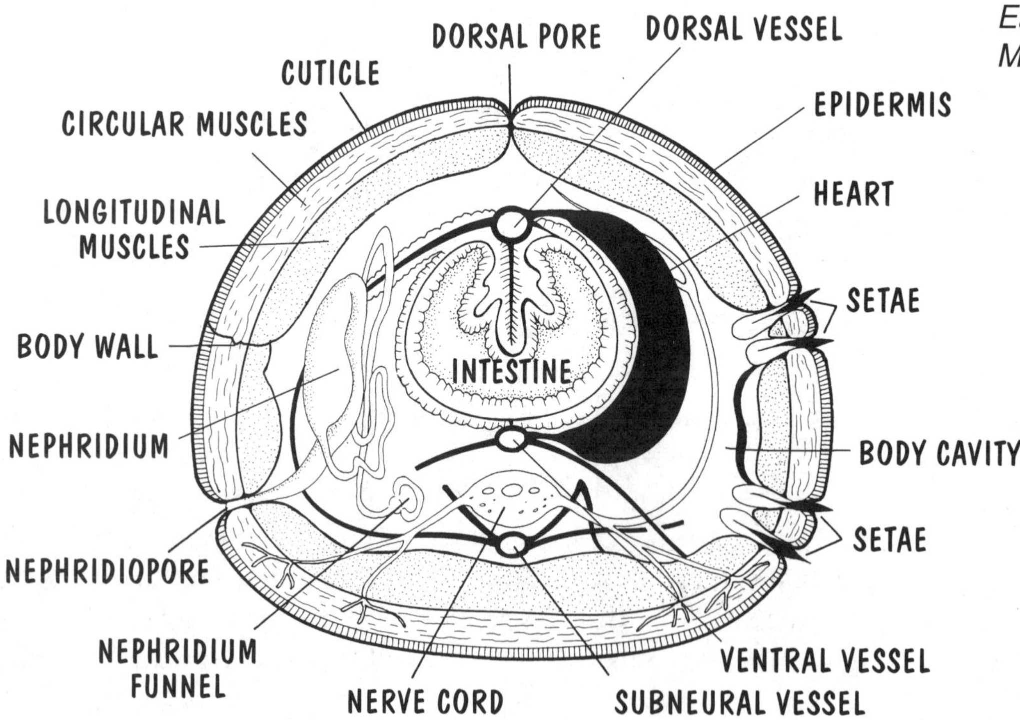

Figure 25. *Earthworm Musculature*

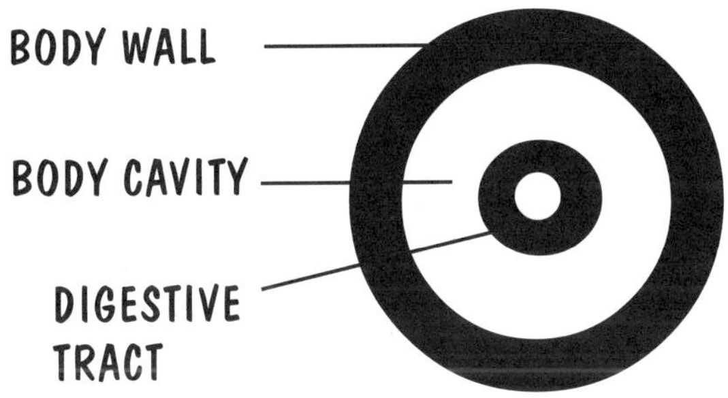

the cuticle and epidermis (*see* Figures 23 and 25) covering a muscular body wall. An inner ring represents the digestive tract. The area between these outer and inner rings is the body cavity which contains other organs.

Body Wall

The body wall is composed of an outer layer of circular muscles. When contracted, these ringlike muscles decrease the diameter of the worm's body. The inner layer of muscles lies longitudinally along many segments; when contracted, these longitudinal muscles shorten the worm's body. An earthworm moves by alternately contracting and relaxing both sets of muscles.

Figure 26.
Internal Organs

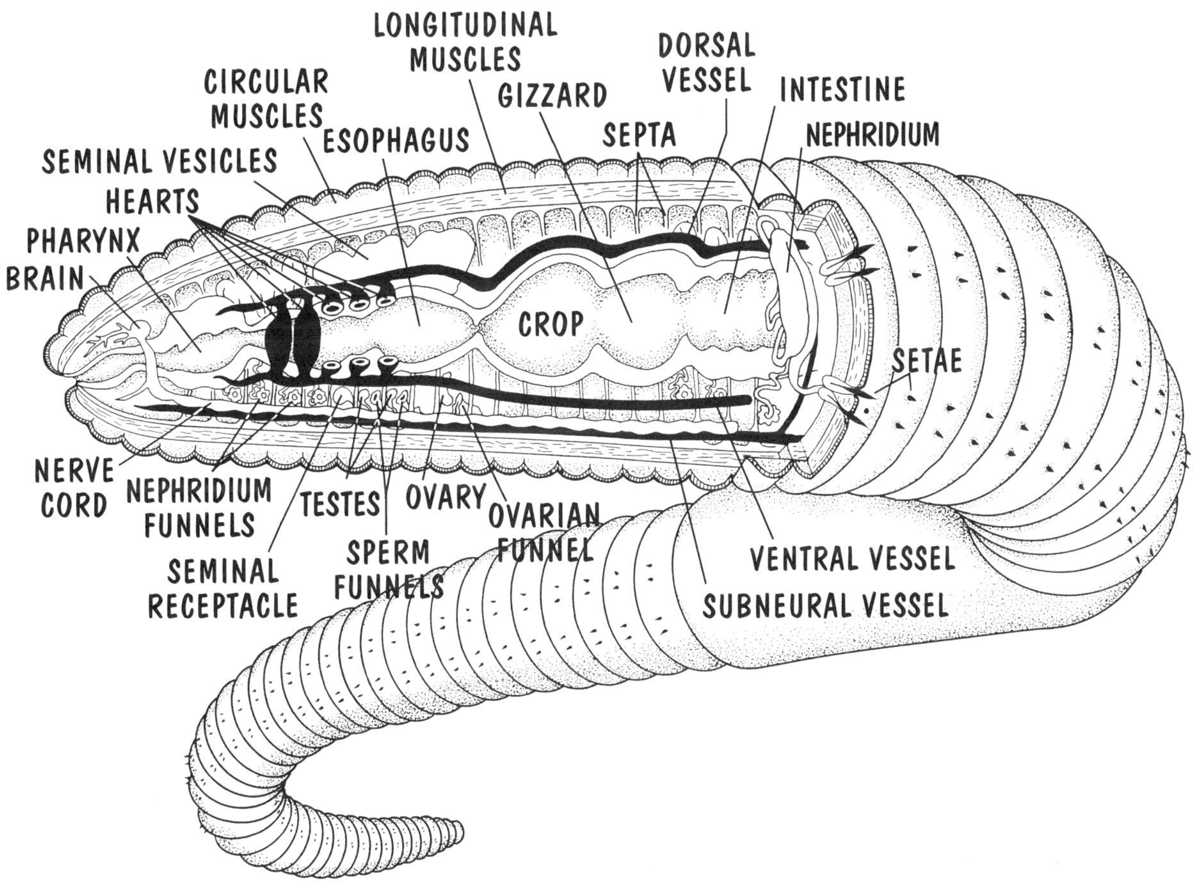

Respiratory System

Water is essential for worm respiration. Since an earthworm does not have lungs, it cannot breathe. Instead, the exchange of carbon dioxide for oxygen (respiration) takes place through the moist surface of the worm's skin. An earth-worm's body contains between 75% and 90% water.

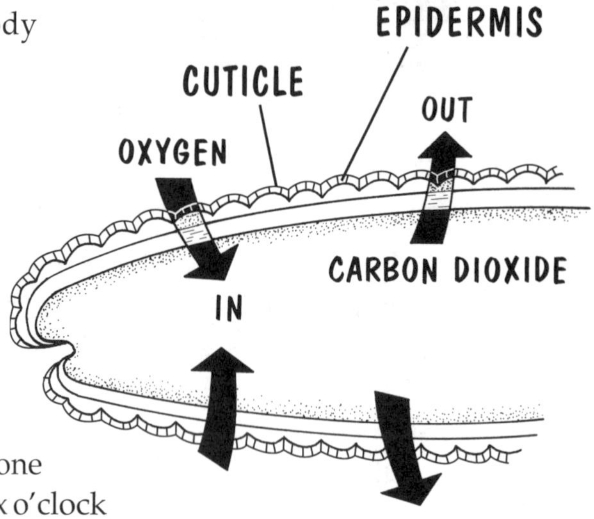

Circulatory System

The circulatory system consists of two main longitudinal vessels. One dorsal vessel lies along the 12 o'clock position and one ventral vessel lies along the six o'clock position, in the body cavity space between the body wall and the digestive tract. Small, curved vertical vessels connect the dorsal and ventral vessels inside each segment. Five arches called hearts, surrounding the esophageal region, pump blood from the dorsal vessel to the ventral vessel (*see* Figures 25 and 26). This circulatory system distributes blood to all other segments. An earthworm's blood contains a hemoglobin-based pigment, not unlike the pigment found in our own blood, which gives the worm its reddish color and carries respiratory gases.

Digestive System

An earthworm's digestive tract (*see* Figures 25 and 26) consists of a three-dimensional lumpy tube that stretches all of the way from its front "head" to its rear "tail" end. Food is taken into the mouth with the aid of the prostomium, a flap or fleshy appendage located above it (*see* Figure 23). The prostomium always reminds me of an elephant's trunk, only much, much tinier. Food passes through the worm's mouth and pharynx, next into the esophagus, and then lands in the crop, a small internal organ, where it is stored until it moves into the gizzard. As a dense muscular organ, the gizzard holds tiny, previously swallowed, particles of rock that grind against the passing food. The food is digested in the intestine (*see* Figures 25 and 26). Indigestible remains become worm manure or castings, as we call them. Castings move posteriorly out of the worm's body through the anus.

Nervous System

The earthworm has a small "brain" (*see* Figure 26) consisting of either a pair or large collection of nerve cells (cerebral ganglia) located above or dorsal to the pharynx of its digestive system. A nerve cord runs from these ganglia to the rear of the worm's body and branches out in each segment to form smaller ganglia (*see* Figures 25 and 26). Sense organs and sensory cells inform the worm about its environment. Messages are sent along nerves that cause the worm to react to stimuli. For example, a bright light stimulus sends a message along the nerve that causes the worm to move away from it. Earthworms are extremely photosensitive and will usually move away from light.

Excretory System

Organs called nephridia or "little kidneys" are also found in each segment (except the first three and the last two). They comprise the excretory system. Nephridia withdraw wastes from the earthworm's blood and interior spaces, and pass them outside of its body through small pores near the base of the ventral setae (*see* Figures 24-26).

Reproductive System

Earthworms are hermaphrodites. Locations and numbers of reproductive organs vary with species, but generally each animal may have two pairs of male organs and one pair of female organs inside segments numbered nine through 16 from the forward end containing the mouth and brain. Most species have a pair of male organs in segments ten and 11, and a pair of female organs in segment 13. The clitellum, also called the "saddle," is a band which appears only on sexually mature worms about 30 segments behind the "head" (*see* Figures 23, 24, and 26).

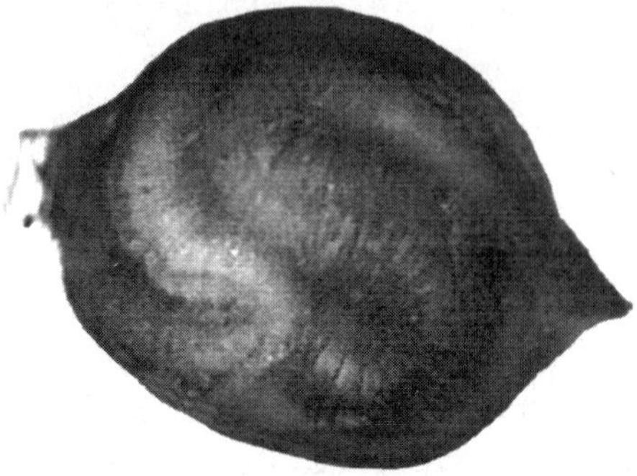

Figure 27.
Redworm Cocoon with Developing Young

Redworms are capable of reproduction at four to six weeks of age, but six weeks is more common. Maturity depends on their environmental conditions. Earthworms mate with their heads pointed in opposite directions. Setae and mucus hold them together. During mating, mucus secretions cover the worms' bodies around the clitellum to the ends of their tails. Sperm are released from the male organs of both worms and deposited into the receptacles of their partners. The worms then separate from each other. At a later time, a clitellum secretion forms a ring and slides forward over the worm's body, gathering eggs from the female oviducts and sperm from the other worm. As the cylinder ring of mucus slides off the worm's head, the ends of the mucus ring close to form a cocoon which holds the fertilized eggs. Each earthworm can produce two or three cocoons per week. In 14-21 days, depending on environmental conditions, two or three worms will hatch from each cocoon.

Reproduction in Worm Bins

How quickly the eggs hatch depends on temperature, the health of the cocoons, and the habitat conditions of the bin into which they are deposited. Both reproduction and population growth are affected by the conditions in the worm bin. Space, food availability, temperature, acidic conditions, and lack of oxygen can all affect the rate of reproduction.

As the earthworm population in the bin grows, additional food waste must be supplied. We found that adding a constant amount of food waste to a worm bin slowed population growth that eventually stabilized. Because the shredded paper waste surrounding the worms as bedding has little nutritive value for them, they become small and puny when deprived of enough substantial food.

Companions on Our Planet

From the inside out, earthworms are fascinating creatures. They are able to consume their own body weight in organic matter each day, depositing in its place the most complete soil amendment available. Quiet, gentle creatures, they tunnel through soil increasing its porosity, structure, and organic content. Each worm has the physical ability to ingest soil and plant material. With the aid of secretions from its digestive system, whatever is ingested is converted to rich castings containing more available phosphorus and potassium than its pre-digested state. In mucus-lined worm burrows is also a concentration of free nitrogen-fixing bacteria.

Soil erosion, today abetted by poor agricultural practices, deforestation, and the use of chemical fertilizers, has denuded the earth of topsoil which decreases beneficial soil organic matter. As reported by the World Watch Institute, in 1990 alone, over 480 billion tons of topsoil was redistributed over the Earth's surface, mostly to unusable locations like river bottoms. Charles Darwin estimated in the nineteenth century that 63,000 worms in an acre of good soil would deposit approximately 18 tons (16t) of castings per year or about two inches (5cm) of new topsoil in ten years.

As animals without backbones, it is ironic that they play such a crucial role in recycling substances on which other, more advanced life forms depend. In a figurative sense, they serve as a backbone for more complicated animals. Without the support and foundation they provide, life around the world would most certainly suffer. Presence of earthworms is an indicator of healthy soil. Their role in the environment is a vital one.

We need to understand and teach the direct connection between organic decomposition and the health of our planet. Whatever we do to the planet, we ultimately do to ourselves. We are all connected in the great web of life:

> ***This we know,***
> ***the earth does not belong to us;***
> ***we belong to the Earth.***
> ***This we know.***
> ***All things are connected***
> ***like the blood which unites one family.***
> ***Whatever befalls the Earth***
> ***befalls the sons and daughters of the Earth.***
> ***We did not weave the web of life;***
> ***we are merely a strand in it.***
> ***Whatever we do to the web,***
> ***we do to ourselves.***
>
> — Attributed to Chief Seattle

Chapter 7

Mid-Scale Vermicomposting as an Ecosystem Model: Don't Sweat the Sow Bugs

If you follow the same normal procedures that Laytonville schools did, you and your students will set up a bin with shredded paper bedding, a starter culture of redworms, and a quantity of soil to provide grit for the worms' gizzards. That soil has also inoculated the bedding with a variety of bacteria to start decomposing the food waste, but after a while — surprise!

Be prepared to enter the "waste frontier," also known as a healthy worm bin community, teaming with an assortment of interesting critters. You may find millipedes with myriad legs moving in waves down their dark elongated bodies. You may find beetles scuttling about in the darkness searching for prey. Closer scrutiny may reveal different beetle varieties. A peek at a reference book on insects shows you that your worm bin has ground beetles and rove beetles in it. Snails may be leaving their moist slime trails as they scrape food with their raspy tongue-like radulas. More likely, you will see slugs, similar to snails, but with their shells greatly reduced to small plates on their backs. You may see squirming enchytraeids or white roundworms.

You are almost certain to see sow bugs busily scampering about feeding on vegetation. Springtails, munching on mold, may hop about if distracted from their luncheon. Mother earwig and her brood hurriedly scoot away from the light. The interesting texture on a rotting tomato turns out to be really thousands upon thousands of minuscule mites that have claimed the tomato their planet. These are the arthropods and other small animals you can see with your naked eye. Here are examples of the many organisms that may be found in worm bins: snails, earwigs, mites, beetles, springtails, milli-

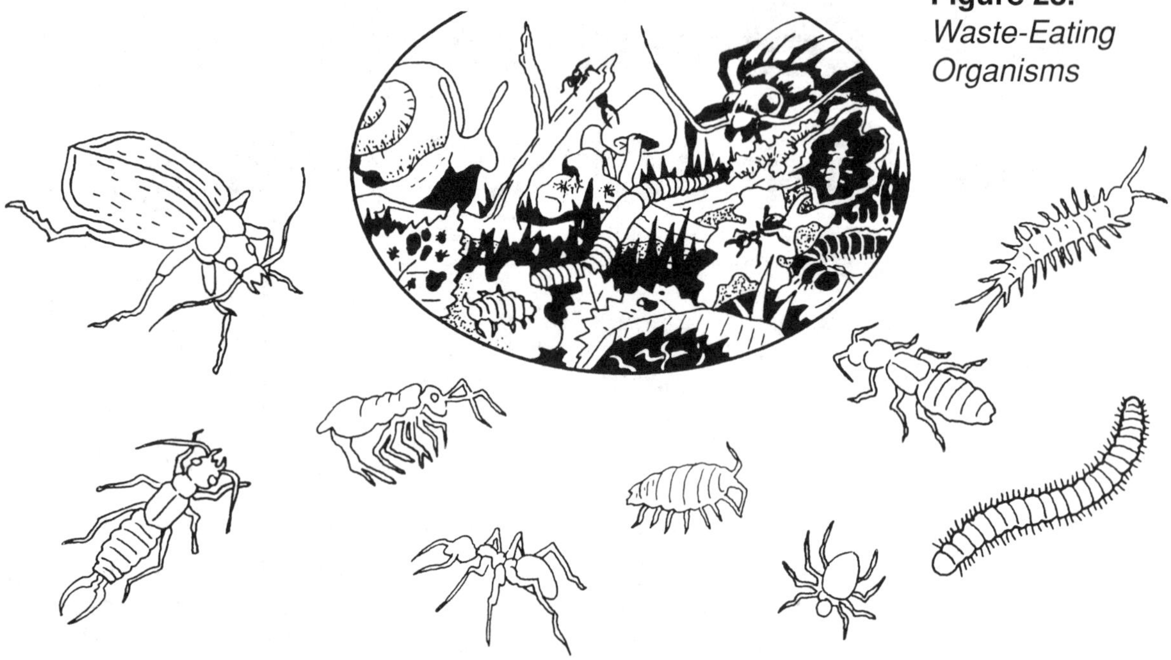

Figure 28. *Waste-Eating Organisms*

pedes, ants, earthworms, nematodes and sow bugs. You will also encounter bacterial colonies and fungi which feed on the different types of vegetable and grain waste.

You may even find a frog or toad nestled down in the moist environment so close to an interesting and available food supply. On inspection with a microscope, you would discover another universe of life. The bacteria, among the first forms of life on our planet, are busy consuming, being consumed, and multiplying.

Where did all of these creatures come from? Have you ever heard of the expression, "Build it and they will come"? Well, you built it and, by golly, they have arrived! They are living in your worm bin, and they are there to do their jobs! These creatures were on the surfaces of the materials you put into your worm bin. They come equipped for combat with an array of technologies for infiltration and decomposition. The message of the day is "Get, before you get got," because the death of one creature is survival for another. What you see is life in its rawest, most basic form. It is the living or biotic part of the worm bin ecosystem.

Your bin is a stable community maintained by constant change, a place where individual bacteria, fungi, and animals interact and exchange energy and materials. The more diversity you have in your bin, the stronger the community. These organisms all have an important part in the complicated system of decomposition. Some feed on each other, and some eat the waste generated by others in the bin. I have never experienced a population of worms being hurt by the variety of creatures which occupy a healthy worm bin community.

Each vermicomposting bin is an ecosystem — a natural system whose patterns of interaction among the decomposers and their predators are unique. When several different kinds of organisms live together and interact with one another in a particular place or habitat, we call their relationships a biological community. We control the physical environment of vermicomposting by providing these elements:

1. *Space*, as defined by the confines of bin construction
2. *Water*, adding moisture as needed
3. *Oxygen*, added by turning bin contents weekly for aeration
4. *Food*, as buried cafeteria wastes

Figure 29.
Controlling the Physical Environment

Thus, each bin ecosystem is an interacting combination of a biological community and its physical environment. One worm bin may have thousands of sow bugs teeming in it, and the bin right next to it will only have a few sow bugs but a million earwigs. Many factors contribute to the population of creatures found in worm bins: location, bacteria, moisture, fungi, foods, and temperature — all play a part in bin populations. Each bin is equally healthy. In the dance of life, each bin community follows a different drummer. For each species, there is a healthy density.

Teaching Ecology in the Schools

Over the years, much in education has changed. Historically, teachers have viewed students as empty vessels to be filled or as future factory workers to be molded by a strict, common standard. The political upheaval of the 1960s highlighted our responsibility for one another, and we actively began to teach social consciousness. But our view of students has sadly remained unchanged. In classrooms across the nation, they are still perceived as empty vessels waiting to be filled with knowledge. That paradigm is shifting only recently.

As teachers, we are beginning to understand that our students come to us as distinctly different, living, growing individuals who bring their own unique experiences to the learning process. As we become facilitators in the learning process, our understanding includes the imperative that each child must construct his own meaning. Information comes both from his own personality and vicariously from others. Learning is a whole-brain, whole-body process. It needs to be relevant, experiential, and connected to action.

Earthworm science students in my classroom create computer-generated graphics, including time lines for earthworm development (*see* Figures 31a-c). They have also used computer technology to create brochures and write stories about their projects. Sometimes they have downloaded illustrations from on-

Figures 30a-c. *Students produce computer-generated graphics like these:*

a.

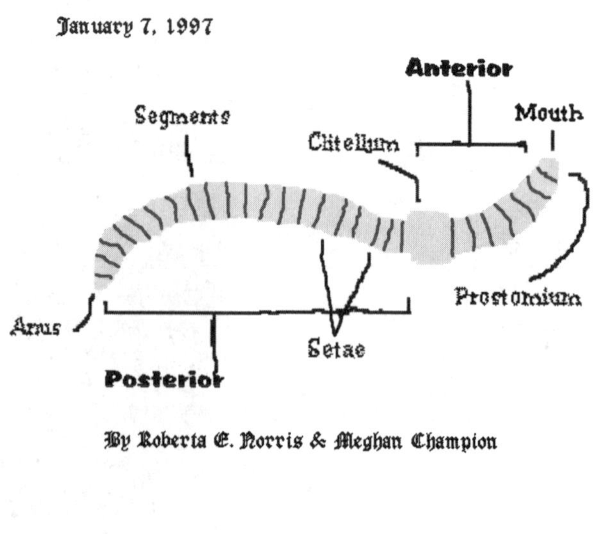

b.

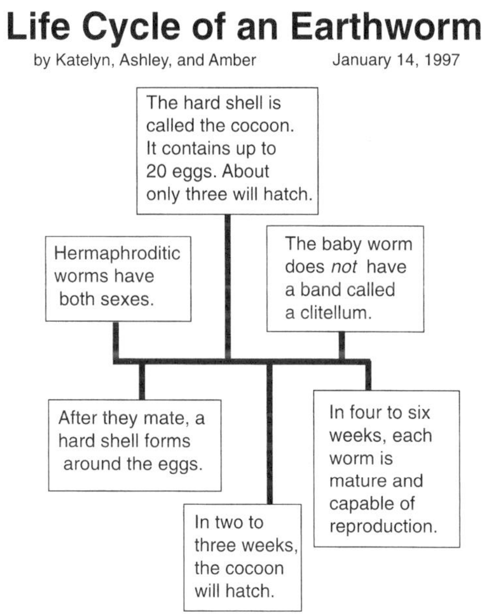

c.

line encyclopedias to use for classroom reports.

Students need to be prepared for a much different world than the one known in the past. Meaningful education for a better world is a priceless intangible. Earthworm science has the ability to reconnect children to the planet and to help them gain an understanding of that interconnectedness. Vermicomposting is a service to one's community both locally and globally.

Today's popular culture includes much talk about the electronic World Wide Web of computer technology. Webs, however, are originally organic in nature. Each linear strand of a spider's web, for example, connects to more than one other strand at many points. Vermicomposting demonstrates the existence of natural food webs in which different species are the connecting points which linearly relate one organism to another.

Model Ecosystems

Energy flows through ecosystems in two interconnected food webs:

1. *Grazing food web,* where herbivores eat plants and are in turn eaten by carnivores and omnivores
2. *Detritus food web,* where the detritus feeders and decomposers break down organic wastes into simple inorganic compounds, which can be utilized by plants

Detritus literally means a product of wearing off or rubbing away, such as decaying leaves and rotting food scraps. Earthworms are included in the community of decomposers known as detrivores. Detrivores feed on dead organic matter or detritus. Animals in this class include crabs, millipedes, ants, termites, and vultures. Scavengers, such as crows and vultures, feed directly on animal carcasses; dung beetles feed on animal manure. Non-animal detrivores include both fungi and bacteria.

Many of the carnivores and omnivores in the above ground grazing food web eat the higher-level predators in

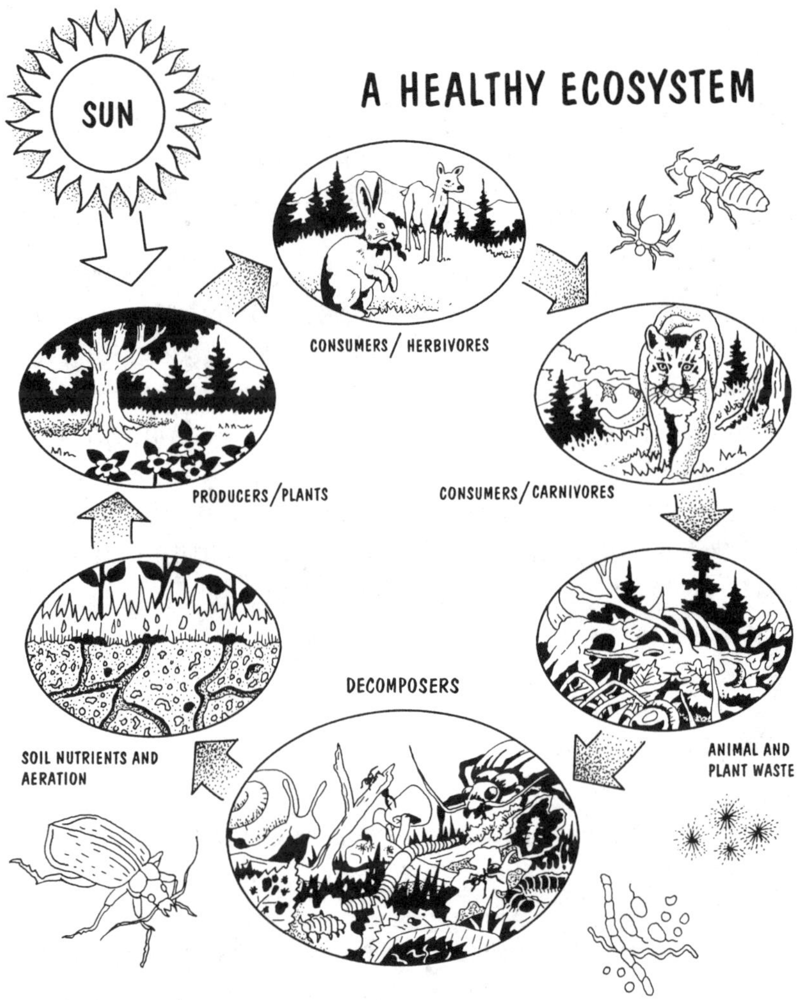

Figure 31. *Energy Flow, Producers and Consumers*

the detritus food web. In fact, the above ground web cannot be maintained without the detritus-based web.

Decomposers are responsible for completing the cycle of life in a healthy ecosystem. Without their help we would soon find ourselves buried in a sea of animal bodies and plant waste. In a healthy functioning ecosystem, there is no waste. Decomposers break down the manures of all animals. The cycle works very simply: the earthworm eats plant waste and the bacteria and fungi that decompose plant detritus; the ovenbird eats the earthworm; and, the bobcat eats the bird. When all of these animals die, decomposers eat their bodies. During consumption of these organic wastes, the decomposers eventually make the nutrients which reside within those bodies available to plants, our primary source of food.

Several levels of decomposers exist. Some are interdependent. Others tolerate one another because they are attracted to the same environment or actively inhibit one another or are completely invisible to one another. Their invisibility makes them neutral. They get along to go along.

Recyclable nutrients ready for use by plants are produced in aerobic compost piles. Fungal and bacterial enzymes break down the organic waste we add. These nutrients are stored in the form of bacterial and fungal biomass. By consuming these bacteria and fungi, the many protozoa, nematodes, and earthworms release nutrients for plant use.

In the worm bin, most of the organisms are decomposers and more than one kind of decomposer is at work. First, the primary decomposers, meaning bacteria and fungi, consume the food waste and are then eaten by fungal-feeding mites and springtails (*see* Figures 28 and 31).

Springtails (Collembola), white-worms (enchytraeids), and flies eat fungi, while worms and non-worm decomposers eat bacteria, protozoa, and other fungi. Worms break up larger plant chunks to make more surfaces available for decomposition by bacteria and fungi. Therefore, worms speed decomposition because they help communities of bacteria and fungi grow faster. As leaves decay and food scraps rot, small pieces break off to become food for organisms that lack the ability to handle larger amounts of food.

Ecosystem discussions usually focus on energy flow and materials cycling more than describing specific plant and animal groups. The original source of energy is the sun's radiation. Plants trap the sun's energy and through photosynthesis combine carbon dioxide from the air with water and minerals from the soil to make food. To living organisms, energy and materials mean making and / or getting food. Energy flows through ecosystems by organisms eating each other.

When a grazing animal (herbivore) eats grass, energy and materials flow from the plant to the animal. Some of the energy is lost in the process through body heat, while part of the energy is used to power the organism. If a wolf eats a lamb, some of the lamb's energy and materials flow to the carnivore and, again, some of the energy is lost. Does energy flow from the carnivore back to the plant? Not likely. Materials can flow back to the plant, but not until other kinds of organisms (decomposers) get involved. This sequence of producer (the plant), herbivore (the lamb), and carnivore (the wolf) is called a food chain. Materials retained by the animal allow it to grow; materials lost by the animal undergo biodegradation into simpler forms ready for recycling into other forms.

Chapter 8

Initial Bin Set Up

Along with constructing your worm bins and reading all the information about beddings and the earthworm's life requirements, your food waste audit (Chapter 2) can help determine how many worms you will use to initially set up your bins. You must also locate a source for the worms.

A rule of thumb is that two pounds (<1kg) of worms can process one pound (<0.5kg) of food waste per day. So the addition of 32 pounds (<15kg) of food waste to a fully stocked bin containing 64 pounds (<30kg) of worms would sustain a healthy population. Using formulas from *Worms Eat My Garbage*, ideally a bin just deep enough to bury the food waste and which measures 32 square feet (<3 square meters) in surface area would support a population of 64 pounds (<30kg) of redworms, or two pounds (<1kg) of worms for every square foot (940 square cm) of surface area.

Figure 32. *Worm:Food-Waste Ratio*

At a price of $10-24 per pound, cost may prohibit beginning with the optimum quantity of worms. Fortunately redworms multiply quickly under favorable conditions, so beginning with even a small population will in time yield plenty of worms to handle your food waste. You can purchase redworms through a local supplier or by mail from growers located in all parts of the country. Look for ads in gardening and fishing magazines.

I began our project with a big, five-gallon (19-L) bucket holding approximately five pounds (>2kg) of worms mixed with some horse manure in one 4x8-foot (>1x2-meter) bin 12 inches (30.5cm) deep. Within ten months, the number of worms had multiplied to keep not only my first bin vermicomposting food waste, but allowed for the addition of three

Figure 33. *Garbage as a Still Life Subject*

more bins. Be assured that the reproductive rate of earthworms, even with small starts, will yield plenty of worms to handle your food waste before long. From our first bin, we processed an impressive amount of cafeteria waste within the first ten months: 3600 pounds (1620kg) of food waste total or 360 pounds (162kg) per month, 90 pounds (40.5kg) per week, or 18 pounds (>8kg) per day.

The more worms you can start with, the sooner you will be able to compost the majority of your non-protein food waste. If you start with eight pounds (<4kg), given worms' rate of reproduction, you should reach optimum stocking rate for each bin at about the 12th or 13th week of operation. My advice is to begin with as many redworms as possible, whenever and wherever you can; don't wait. The bottom line is to start with one bin and as many worms as you can afford.

I used Dr. Roy Hartenstein's laboratory studies on earthworm reproduction to determine the rate of population growth. With an outdoor bin, you will not have as much control as Hartenstein had in his lab, but it will be close enough to approximate the same rate of growth depending on water content, weather, and the living conditions of the worms. Optimum stocking rate would be two pounds (<1kg) of worms per 1 square foot (<0.5 square meter); minimum, whatever you can afford. Just remember that a small worm population will not process a high volume of food waste. Under average conditions, the 2:1 worm:food-waste ratio will provide enough worms to compost all of the food waste you generate.

Initial bin set up requires the following materials:

1. *Shredded paper* or some other bedding provides the medium in which to bury the food waste.
2. *Water*, added to the bedding at roughly three times the weight of the bedding, facilitates worm respiration. The evaporation of the water in the bedding during the summer months can keep your worms happy on a hot day. Always be sure to locate your bins near a source of water.
3. *One standard shovel of soil*, about a half gallon (>2L), provides grit for the earthworms. The soil also introduces bacteria into the bin.
4. *Redworms*, enough to start a mid-scale bin would weigh approximately two pounds (<1kg) or more, are needed to produce castings by the digestive process.
5. *Food waste*, at half the weight of the available worms per day, should be added to supply energy and materials for your worms.
6. *One* 4'x8' (122cm x 244cm) *sheet of black plastic* will cover the contents of the bin to help control light and surface moisture.
7. *A sturdy bin* (*see* Chapter 3) is necessary to hold materials numbered 1-6.

I always mix the bedding with water directly in the bin before adding the worms. An earthworm's body is between 75%-90% water. Dry bedding will injure their delicate tissues and suck the life right out of them. A good rule of thumb is to use three pounds (>1kg) of water to one pound (<0.5kg) of bedding. As a child I learned that a pint (<0.5 liters) is a pound the world around and I have always passed this mnemonic on to my students. Three pints of water (1.5 quarts or >1 liter) weighs about three pounds (>1kg); little difference exists between these English and metric units.

Eventually the paper bedding should be damp, but not soggy. If anyone can squeeze water out of the paper, it is too wet. Beddings which are consistently too wet will not allow the worms to breathe; suffocation from lack of oxygen will claim them.

At the beginning of the project, we weighed both the paper and water. As we became more familiar with the food waste and paper flow, we discovered weighing them was a time-consuming step we could skip. By thoroughly mixing the water with the bedding beforehand, excess moisture will have the time to drain through the holes located in the bottom of the bin before the worms are added. Because our bins are located outdoors, excess water drains onto the ground and

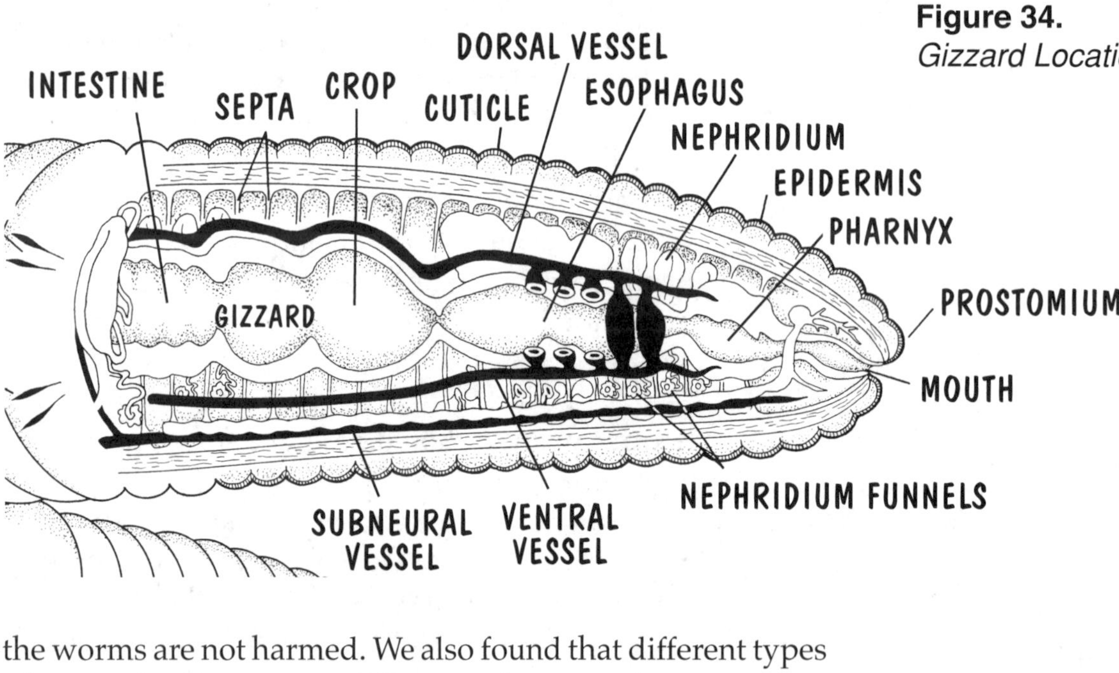

Figure 34.
Gizzard Location

the worms are not harmed. We also found that different types of paper absorb water at different volumes and rates.

Along with the moistened bedding, a relatively small amount of soil is a necessary addition. Soil particles in the earthworm's gizzard (*see* Figure 34) act as grit against which ingested food is ground. After adding the soil you are ready to inoculate the bedding with worms.

Chapter 9

Working Your Bins

If you are starting with a small worm population and one bin, carefully monitor the food waste. Add food at a rate not to overload the trench system described below to bury the food waste. After the worms are added, the bedding is ready to receive food waste in two phases.

Phase One

Divide the bedding in each bin into eight equal sections, from front and back along its length. Assign each section with an alphabetical letter. If you do not have enough food waste to fill a twelve-inch trench in every section each day, you can divide your boxes into 12 or more sections and make narrower trenches, but remember to use the same principles. More trenches will give the worms extra time to work on the food waste, and your vermicompost will be more finished when you remove it from the bins.

After filling a trench, always first cover it with dampened paper bedding and then cover the entire surface of the bin with a large piece of black plastic. Besides helping to keep moisture in, this dark cover shields the worms from natural light. Earthworms are extremely photosensitive and will always move away from light sources. Bait catchers searching for nightcrawlers with strong flashlight beams know how quickly worms react to even artificial light. After the plastic is in place, cover the bin with its plywood lid. The plywood protects the bin from excessive moisture when it rains. It also keeps rodents and other wild creatures from helping themselves to the worms and food waste. Bears have been known to be partial to earthworm bins in our area.

Week 1, Day 1. Dig a trench in the first section (A, *see* Figure 35) of the bedding, and bury the food waste evenly into the trench. Make sure you completely cover the food waste with dampened paper bedding so it composts more quickly and does not attract fruit flies. If a fruit fly infestation occurs or if you notice foul odors at any time, withhold food from the bin until the balance is restored.

You will need to add more and more food waste as the population increases. If you continue adding small or moderate amounts of food at a constant rate, the population will become fairly stable since that constancy will only support the existing population of worms.

Week 1, Day 2. On the second day, skip the second section (B) and dig a trench in the third section (C) to bury that day's food waste. Continue to bury food waste in every other section of the bin, moving from bin to bin if you have more than one. When bin #2 is full, begin adding food waste to bin #3 and follow the same trench rotation as you did in bins #1 and #2.

Depending on the amount of food waste generated each day, it may take a week to bury enough food waste to fill the first worm bin. Then move on to the second worm bin and begin the same burying process again. Do not add new food waste to the first bin until you have rotated through all of the bins. If you only have one bin, you must monitor the worm: food-waste ratio and make sure you are not feeding the worms more food waste than they can process. Just remember to increase the food waste gradually so the population will continue to grow. Observation is crucial.

Week 2, Day 1. On the first day of the second week, go back and turn the food waste in section A which was buried on the first day of Week One. Remember, frequent turning encourages the food waste to break down more quickly because it breaks up the food waste and allows more bacteria to penetrate the food.

Figure 35.
Rotation of Food Waste

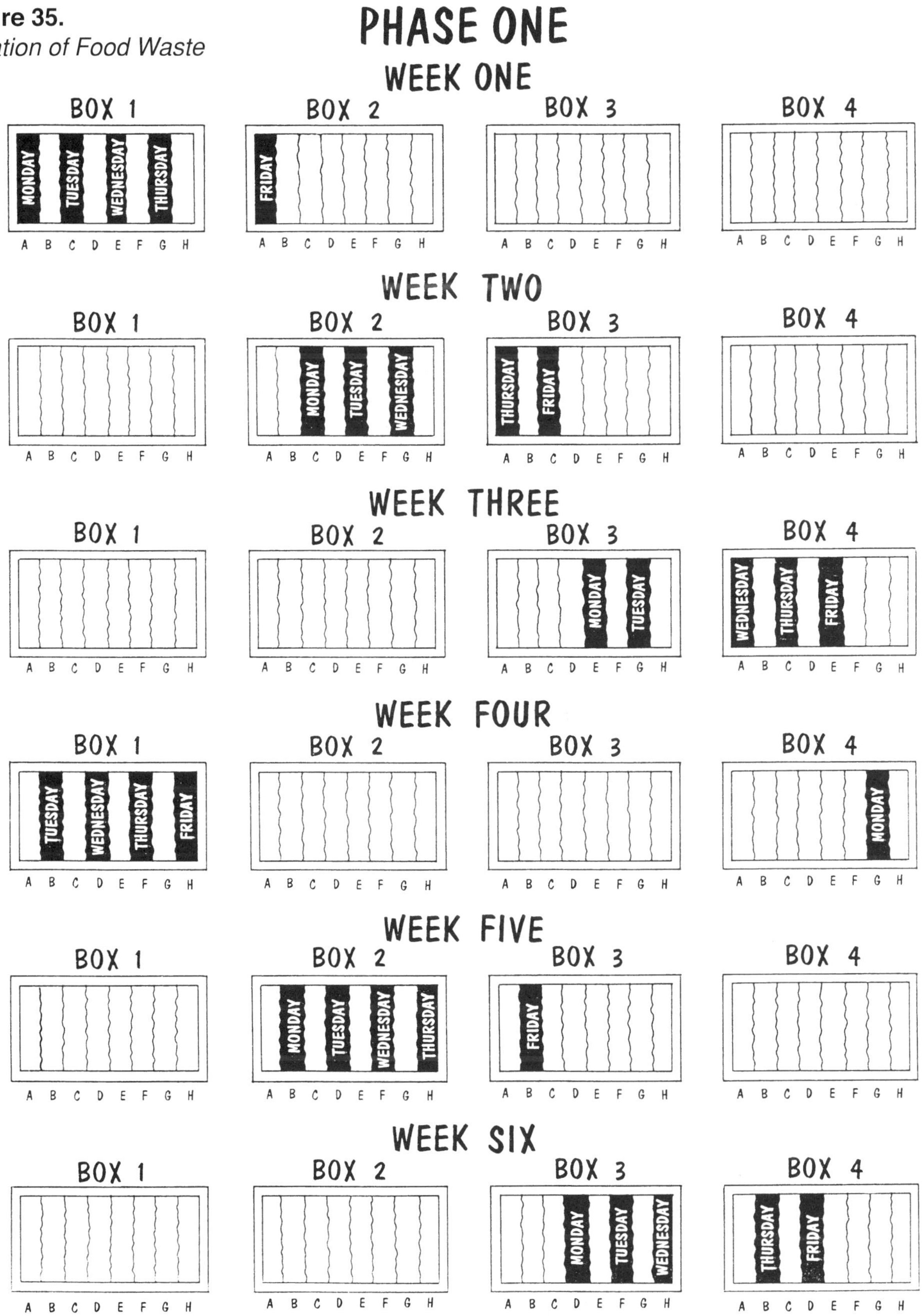

Each day count back in ten-section increments and turn the material in those trenches. Each trench needs to be turned at least once a week. After three weeks, assuming you will eventually have four bins, you will be back again at bin #1. In this rotation, you will use the sections in the bin that you skipped the first time around. You will fill sections B,D,F and H. You will continue adding food waste until all of the sections in each bin have been filled. This will take roughly six weeks if there are no holidays or vacations.

Phase Two

Week 7, Days 1 and 2. Move all of the material, bedding, castings, and food waste of bin #1 into sections A and B (*see* Figure 36). This material will be from two to five weeks old and may still contain some fairly fresh food waste.

Put new bedding into the remainder of the bin and divide it into six equal sections. Begin filling the bins with food-waste trenches as you did before, skipping every other section. Continue to turn the older material in sections A and B of bin #1. As the food waste is depleted in sections A and B, the worms will begin migrating away from their castings and into the fresh food waste in the other sections of the bin.

Week 12, Days 1 and 2. At the beginning of the 12th week (again, if there are no holidays), you will be back to bin #1. This time you will need to remove the castings that you collected in the first two sections, A and B, five weeks ago. This vermicompost will be from seven to eleven weeks old. How much food waste still needs to be broken down and consumed depends on weather conditions, turning, and the population of the worms in your bin. Again, turning the finishing compost frequently encourages faster decomposition. If the vermicompost is well processed, it can be used immediately. If there is still a small amount of food waste in it or if you do not need to use the vermicompost at once, move the vermicompost to a "finishing box" where it can be used as needed. Leave the rest to finish composting.

I always use a finishing box: a bin for holding vermicompost while it continues to be broken down (finished) by the microorganisms in the bin. My students screen the vermi-

Figure 36.
Rotation of Food Waste

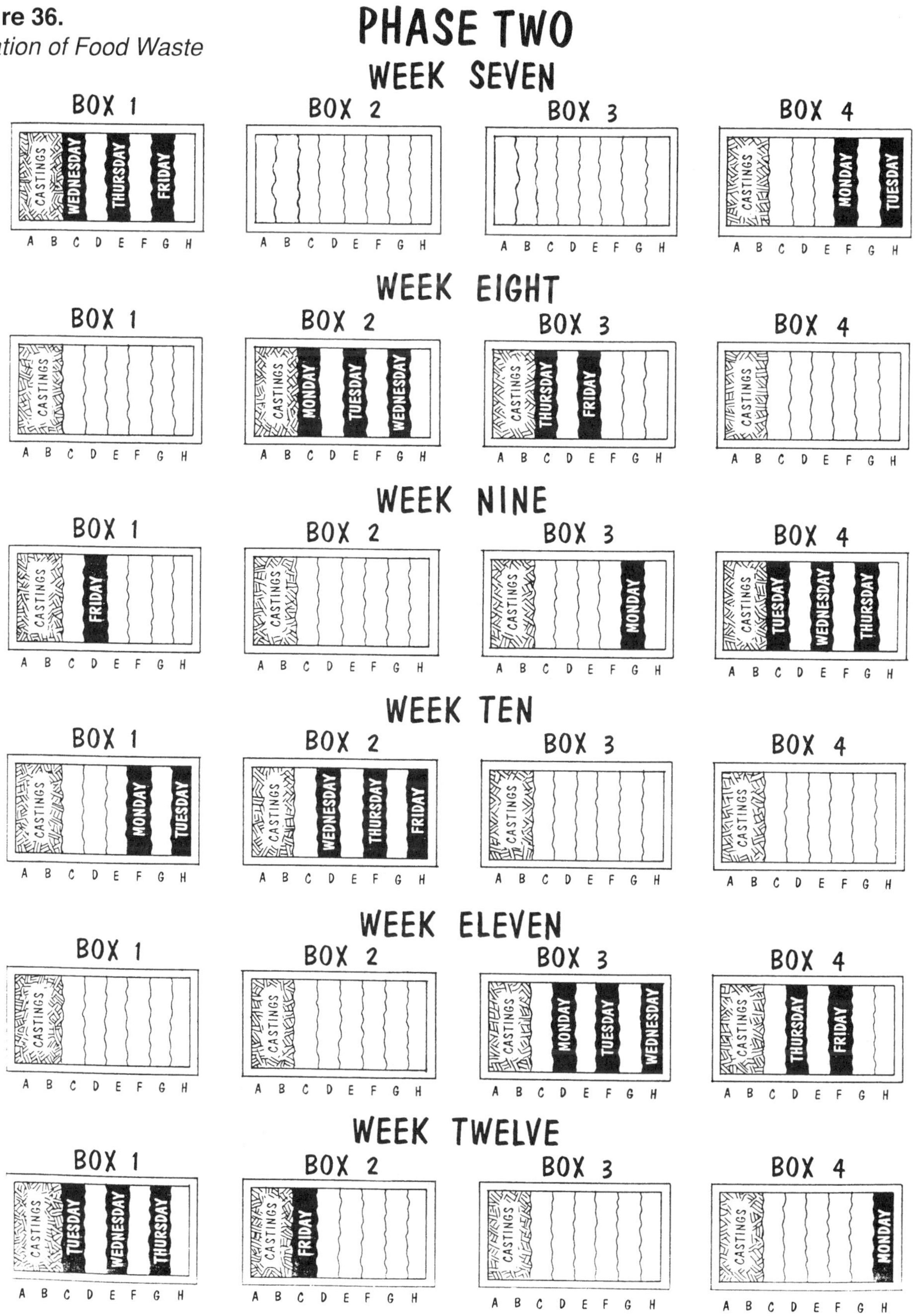

compost through half-inch hardware cloth to remove any food waste which needs more time to break down. The food waste

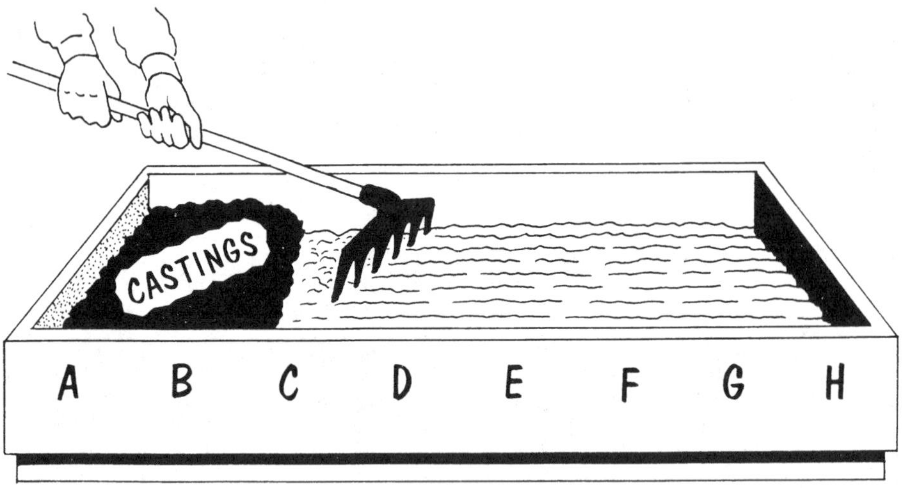

Figure 37. *Collecting Castings*

is returned to the finishing box where the worms and decomposers continue to break it down. This box, shown later in Chapter 12, also provides a location in which to hold or store the castings. It is difficult to take castings directly from the

Figure 38. *Removing Castings from Working Bins for Transport to Finishing Box*

working bins because they often still contain small amounts of food waste. Because our vermicompost is used by many teachers and their classes, we must have an area where teachers can just come and get the vermicompost for their gardening projects without disturbing the working bins.

After the finishing vermicompost in sections A and B is removed, scrape everything left in the bin into sections A and B, add new bedding, and begin burying food waste again. Continue through all of the bins, mounding the partially composted materials into sections A and B of the bin, add more bedding, and make new trenches as needed.

Week 16, Days 4 and 5. Now you are ready to remove the vermicompost from bin #1 again. After removing the vermicompost to the finishing box, add new bedding and start the whole process again.

Underneath the plastic sheet and plywood lid of the bin, earthworms quietly go about the business of vermicomposting, taking your school's food and paper waste and converting it into nutrient-rich, useable, saleable vermicompost.

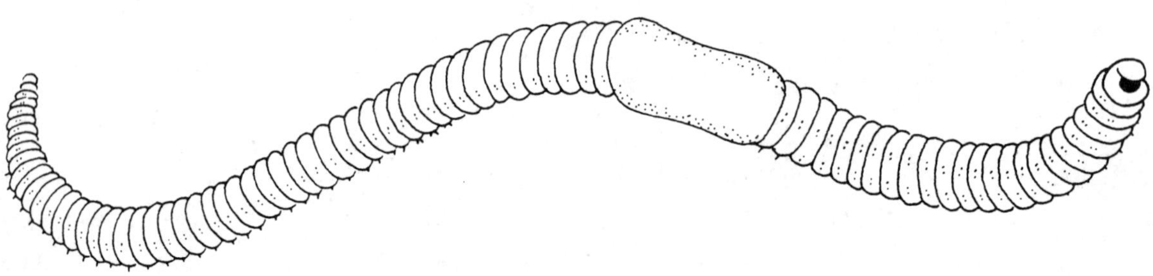

Chapter 10

Managing Food Waste Flow at School

Our food waste is collected everyday in plastic containers located in the cafeteria from both the lunchroom and kitchen preparation areas. Each day after lunch, students pick up the containers and carry them to scales located near the vermicomposting bins also known as worm boxes (*see* Figure 39).

Figure 39.
Layout of Our School Site

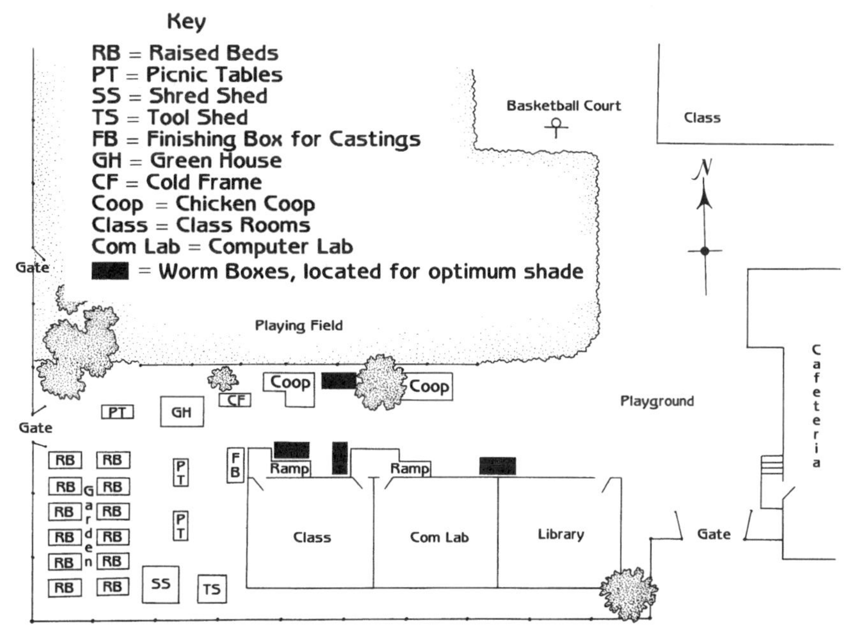

Because I teach math and science, the daily weighing and data collecting fit very well into my curriculum. Fortunately, I am not the only teacher at my site who has worm boxes outside the door of her classroom. Often the food waste does not make it to the big mid-scale bins, but is pre-empted by portable mini-bins used inside classrooms for worm science. Teachers who have small bins, 1 ft. x 2 ft. x 3 ft. or (30x60x90cm) containers proven for home use but here located either inside or outside of individual classrooms, often compost food waste left from snacks in their classroom bins.

Only food waste generated in the cafeteria is weighed, recorded, and kept as data on a regular basis. There have been times when classes have asked to include their classroom food-waste weight data with ours, and we have accommodated their requests. Teachers with indoor mini-bins will ask us to

Figure 40.
Making Indoor Worm Bins

hold the cafeteria food waste in containers for eventual use in their indoor classroom bins. Believe it or not, there are certain times of year when food waste is a negotiable commodity at our school.

How much food waste is added to each bin everyday will be determined by the amount of discarded food waste the cafeteria generates and by the weight of worms in your bins. When we have an excess of food waste and the worms cannot keep up, we turn to other methods of composting. A compost pile or one of many traditional compost bins (*see The*

Complete Book of Compost, by Robert Francis, 1998) designed to handle food waste can be used until you can again add your food waste to the worm bins. When the worm population increases, a bin of the dimensions described in this text can easily process 25 pounds (>11kg) of food waste per day.

I do not recommend storing food waste because stored food waste begins to decompose quickly. It is important for the students to handle the food waste before it begins to decompose and becomes smelly. Students will gradually begin to look at fresh food waste as "food that was not eaten" instead of "garbage." Rarely have I had students who protest the handling of fresh food waste. Because one goal of a school vermicomposting program is to have the students responsible for all of its aspects, their attitudes must be considered.

Laytonville students take daily food waste to the school garden where a hanging scale is located. Scales can just as easily be located in the school kitchen or in classrooms. Regardless of where scales are kept, it is students who use them. They weigh the food waste, and then chart the weight on the appropriate clipboard which at our site hangs inside the door of my classroom.

The purpose of the clipboards is for record keeping. Each bin has its own corresponding clipboard. Attached to each clipboard is a sheet of paper for recording pertinent information (as an example, *see* Figure 41). Students record the weight and location of the food being buried, as well as general daily observations such as problems with fruit flies or rodents.

Figure 41.
A Typical Clipboard Record Sheet

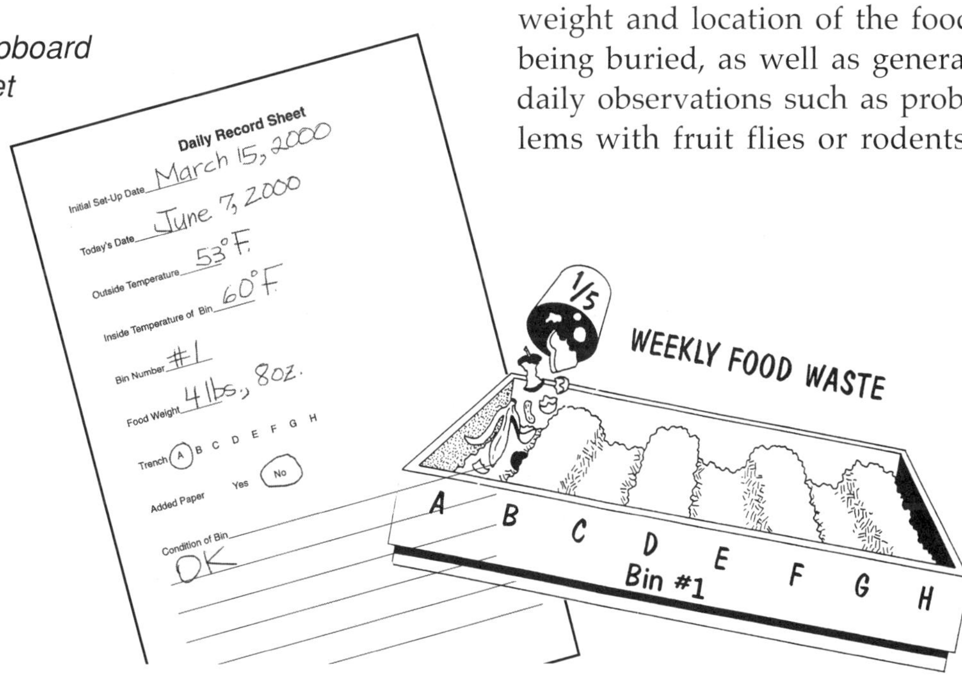

These student observations help maintain the health of each worm bin and insure maximum waste decomposition.

Good record keeping is essential to the success of the program. It is also part of being a good scientist. Recording the weight of the food waste generated each day allows us to report accurately just how much food was vermicomposted and to calculate dumpster fee savings by the district at the end of the school year .

After data are recorded, students add the food waste to the worm bins. They break whole and unbroken food waste, like oranges and apples, with a shovel to hasten decomposition. Each of the four trenches across the width of our bins can hold about eight pounds (<4kg) of food waste, or 32 pounds (>15kg) for a single 4'x8' (>1x>2m) bin per week. These amounts compare favorably with the formula recommended in *Worms Eat My Garbage* of one pound (<0.5kg) of food waste per week for each square foot (900sq cm) of surface area in small, home-sized vermicomposting bins. When students turn the contents throughout the bin with potato forks, or sometimes pitchforks, they effect maximum decomposition. Students then wash the compost containers and return them to the cafeteria for use the next day.

We reduced the volume of our food waste by providing a collection basket in the cafeteria labeled, "Food To Share." Students who do not want to eat their lunches, especially whole pieces of fruit, can put the food into the basket instead of into the trash. This basket allows other students to help themselves to food that otherwise would have been thrown away.

Food waste can flow through a school site efficiently and in a timely fashion each day. Because it is repeated day after day, year after year, vermicomposting becomes routine. The students remember even when I forget. That, my colleagues, is essential to managing an effective program. Everyone is part of the project, knows its details, and works to achieve successful results. The project does not fail because more than a select few understand the what, why, and how of its every aspect.

The mid-scale vermicomposting program at Laytonville Elementary and Middle Schools is very simply managed by children with the aid of adults. All of the cafeteria food waste containers and the design of the bins were created with the idea that the program must be manageable by children. No container must be too large in size or too heavy for students to lift comfortably. Each year, older students teach younger students how to weigh, record, and rotate the food waste in the vermicomposting bins.

Assistance from adults may come from lunchroom staff who help monitor the separation of food waste, as well as teachers being available if students have questions regarding the health of the vermicompost bins. Every one or two weeks I check the records and inspect each bin to make sure things are running smoothly. Temperatures inside the bins, insect populations, and general worm health are all monitored. All classes recycle paper in their classroom, and are responsible for bringing it to a central location for shredding and later use as bedding. Students are taught first to think critically and solve the problems they encounter, before going to an adult for assistance.

So what happens during the summer when there is no school? Worms are fed only one or two times all summer. Recently our school has started a summer program. Soon the bins will get more attention during the summer months, but we've had no survival problems during even long school holidays.

Plan for Overcapacity

Plan to add more vermicomposting bins than will be needed for the initial set up of a mid-scale program because the bins are always in a state of flux. Certain sections of the bins will be out of commission while the contents are "finishing" or while food waste is still being broken down and consumed. Some of the bins will still be working toward an ideal worm population of 64 pounds (<29kg), and some of the worms will be busy migrating to new bedding. Worm migration always precedes the harvesting of the castings.

Each bin should be able to compost 32 pounds (>15kg) of food waste per week, and 128 pounds (<58kg) per month. For that to be true, each bed must have 64 pounds (<29kg) of redworms in it and all must be working bins.

During the first school year of our program, excess food waste was composted in outdoor compost piles until the worms were capable of handling the waste flow. We processed 3600 pounds (1636kg) that year. In our second year, with all bins working, we vermicomposted 2300 pounds (1045kg) of food waste or 230 pounds (105kg) per month, 57.5 pounds (<26kg) per week, 11.5 pounds (>5kg) per day.

I am not sure why we vermicomposted less food waste the second year. Students analyzed the data and suggested the difference was possibly due to a heightened awareness of how much food waste we actually produced. Perhaps knowing made them more careful about waste, and they consciously produced less. By the second year the students were growing produce for the school salad bar and were involved in the cafeteria as well. There was also a small decrease in student

Figure 42. *Vermicomposting Production Graph*

population. For whatever reason, we continue to vermicompost about 2200 pounds (1000kg) of food waste in a ten-month period.

Today, our worm bins each continue to process about 25 pounds (>11kg) of food waste each day. The worms keep reproducing so that we can supply the people of our community with worms for vermicomposting at home and also sell them to a local nursery. Occasionally we have excess worms that we can share. In fact, our school gives away free earthworms to any school who can pay a visit to our site.

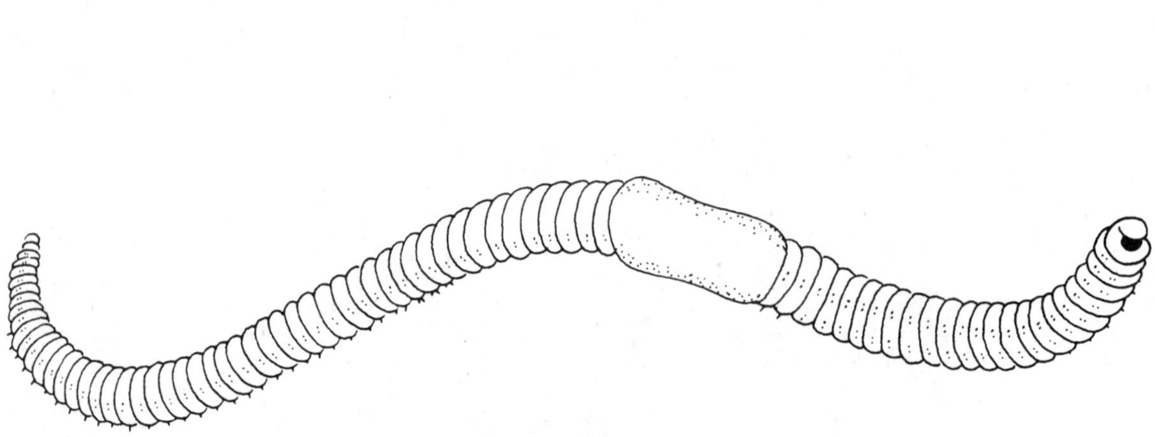

Chapter 11

Managing Food-Waste Conversion Inside the Bins

Each day the collected food waste is cycled through the worm bins. Students dig a one-foot (30.5cm) wide trench about eight inches (20cm) deep across the four-foot (>1-meter) width of each bin. Students then pour the food waste into these shallowly dug trenches and cover it with the moistened paper bedding.

Figure 43.
Cultivating or Worm Forks

I suggest students use cultivating forks to pull back the paper bedding while digging trenches. We call our cultivating forks "worm forks" because they are used for the purpose of working in the bins. They are kinder to the worms than using trowels or even potato forks, which often cut the worms in half.

Cover Food Waste with Bedding

It is important that the students leave no food waste on the surface of the bin for two reasons. First, buried food waste decomposes faster than unburied food waste because more surfaces are directly exposed to the millions of decomposing bacteria in the bin.

More importantly, covering food waste with bedding discourages flies. The odor of food waste attracts various types of flies, including fruit flies and house flies. When food waste is left on the surface of the bin, flies lay their eggs in it. The eggs quickly hatch into larvae (maggots). These feed on the waste, form pupae, then hatch as adult flies. For most of us, this is an unpleasant invasion. Burying the waste underneath bedding contains the odors so that flies are less likely to find the bin, and those that do can't get down to the food to lay their eggs. Preventing fly problems is easier than curing them.

Figure 44. *Covering Food Waste*

Many ways have been tried to prevent fly problems. Possibilities include opening the lid on a windy day, washing out larvae from the bins, and using ingenious traps such as sticky traps or the fruit fly traps described in *Worms Eat My Garbage.*

Monitor Food Waste Consumption by Worms

Initially, students need to monitor the bins very carefully to ensure that no more food waste is added than the worms can eat. If it appears that the food waste is being consumed and processed by the worms at the same rate that food waste is produced, their numbers will increase and they will be able to handle more and more food waste in gradually increasing amounts. If the food waste is not being consumed by the earthworms at nearly the same rate it is being added to the bin, an imbalance of your bin community occurs and problems result. To correct imbalance, slow down the rate at which food waste goes to the bin.

Problems with Excess Food Waste

Excess food waste will cause several problems in your bin. Most noticeable will be a bad smell caused by the production of foul gas created by anaerobic bacteria. Another sign of excess food is the presence of fruit flies. You can do three things to correct these problems:

1. Withhold fresh food from the bin for one week.
2. Add fresh bedding and mix the contents well.
3. Spread a three-inch (<8 cm) deep layer of fresh paper bedding over the entire surface of the bin.

Aerate the Bins Frequently

Once a week turn the food waste and paper in all of the bins for these interdependent outcomes:

1. Encourage good oxygen circulation throughout each bin.
2. Make sure no pockets of anaerobic bacteria are at work.
3. Hasten efficient vermicomposting of food waste.

In our outdoor bins we found that frequent turning allowed us to harvest worm castings much sooner than just letting the bins sit. With turning, good castings can be ready to harvest in two to three months instead of four to six months without turning the contents. Frequent turning also hastens the decomposition process by further bruising the food waste and supplying more oxygen for the worms and the aerobic bacteria present.

Increase of aerobic bacteria, introduced by turning bedding and circulating air, slowed reproduction of anaerobic bacteria. Turning helps air move back into the bedding, allowing aerobic bacteria to best the competition against the anaerobic bacteria. Aerobic bacteria will consume food waste more completely and with less odor.

Frequent turning keeps every organism in the bin working up to its full vermicomposting potential. It also keeps the

Figure 45.
Different Forks for Different Functions

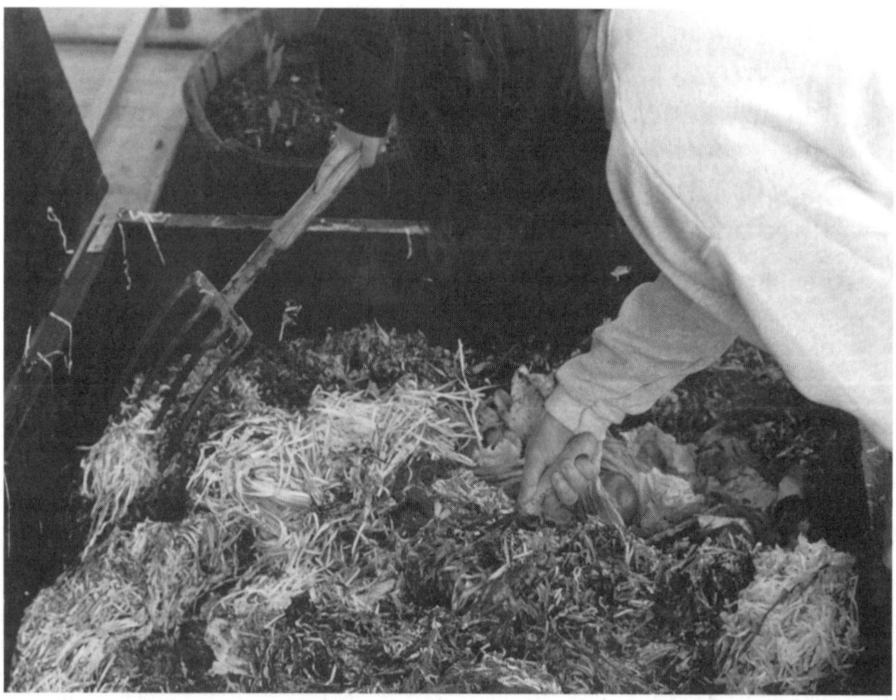

layers of paper from compacting and becoming difficult for the students to turn.

Appearance of Castings

When you first begin it will take four to six weeks before conversion of the paper and food waste to castings becomes noticeable. As the composting community in the worm bin begins to break down the paper and food waste, the volume of the paper in the box decreases dramatically. The bulk of the worm mass decreases when more and more castings fill the bin. As the bin fills with castings, its environment gradually becomes unhealthy for the worms. When you begin to notice more castings than paper or food, the volume should be such that you can now move all the vermicompost to one side of the bin to begin the finishing process. Exactly when this may occur depends largely on the type of bedding used, the average daily temperature, the quantity of worms being used, and the moisture content of the bedding. Moist bedding decomposes faster than drier bedding. However, even though bacteria love a moist environment, too much water will slow decomposition.

Compost Excess Waste

Eventually the worm population in your bins should become large enough to manage all of your nonprotein food waste. When a bin has reached its maximum food waste capacity, it will be difficult to find an area within it where you can bury new waste and it may or may not have a fruit fly problem. These conditions are good indicators of excess food waste.

When our bins reach their maximum vermicomposting potential and cannot take any more food waste until they stabilize, we withhold food waste from those bins. We then put our excess food waste into aerobic compost piles in our school garden. Not all schools will have that option. Even if vermicomposting cannot handle all of your school's food waste, any food waste composted by any method helps reserve our landfills for other discards and reduces use of water and fossil fuels.

At our school site food waste is at a premium because so many teachers use the school garden to teach science. When I have excess food waste it goes to another classroom for the study of soil science through composting. Just remember that every little bit helps and that source reduction done on-site has a direct impact on the quality of life in your community and communities beyond.

Vermicomposting Variables

When I first began mid-scale vermicomposting, like most scientists, I worked hard to control all of the variables (*see* Figure 46). I soon discovered that controlling variables in this

Figure 46. *Vermicomposting Variables*

Variables	Results			Controls
	From Too Much Variable	From Too Little Variable	From Wrong Mix of Variable	
Temperature	Baked worms	Frozen worms		Shade cloth; Insulation
Water	Drowned worms	Dried out worms	Inconsistent dampness	Drainage; Add water
Food Waste	Stunted worms	Starved worms	Bin stinks; Fruit flies	Monitor additions
Bedding	Compaction	Can't bury garbage	Toxic to worms	Monitor additions
Grit	Bin too heavy	Gizzards impaired		Medium standards
Worms	Stunted and starved worms	Bin stinks	Species not compatible	Limit species
Non-Worm Organisms	No worms	Shortage of decomposers	Dead worms	Monitor additions
Children	Boredom, not enough to do	Frustration, too much to do	Rivalry and competition	Schedules
Supervision	Students lose interest	Chaotic project	Wrong ownership of project	Do unto others
Bins	Little room for anything else; Wasted space	Higher disposal costs	Safety hazard; Inefficient; Ugly	Worm growth and weight
Shred Shed	Waste of resources; Higher overhead	Junk all over room, yard		Plan ahead

situation was an impossible task. Daytime temperatures dropped and the surface of the bedding froze. Initially surface feeding was disrupted, but I discovered that worms moved below the frost and continued to feed. One day meat was added by mistake and maggots erupted, but we continued to turn the maggoty mess and added more paper and water. We turned the maggot-infested bin frequently, introducing more and more oxygen in an attempt to salvage the bin. Within a week, the fly larvae moved out and away from the moist, wet conditions inside of the bin to form pupae. Once again the bin began to work without maggots.

Throughout all of our trials and errors we found that, unlike smaller vermicomposting systems, mid-scale systems are more tolerant of a wider variety of conditions that in smaller systems can cause major problems.

Make peace with this bit of wisdom: no one can control all of the encountered variables when weather, waste, worms, and children are mixed. Instead, an economy of scale bestows a saving grace. Since mid-scale systems are forgiving to a point, they thrive when the requirements for any healthy ecosystem are met. I have found no better vehicle for teaching about ecosystems than a mid-scale vermicomposting bin.

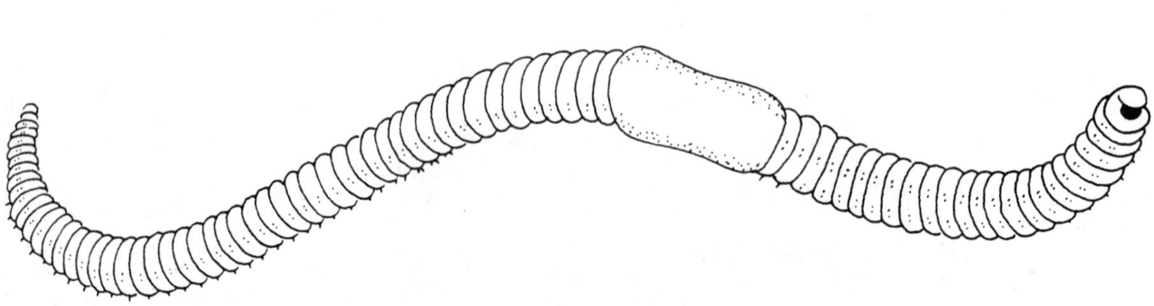

Chapter 12

Top Harvesting Castings

Over the years I have tried many different methods of harvesting finished castings from our worm bins. I began using the "divide and dump" method, an efficient but time-consuming method of hand-sorting described in *Worms Eat My Garbage*. I also tried the "migration method," moving all the almost finished vermicompost to one side of the bin, adding new paper and food waste to the empty side, and letting the worms migrate to the other side with new bedding and fresh food. My students experimented with different foods to coax the worms to the fresh bedding side of the bin. One year, we tried to sift the castings through large screens leaving the worms behind. We were not very successful because the compost was so damp.

The easiest, quickest, and most effective method I have found to harvest castings in mid-scale vermicomposting is a combination of migration and a method I call "top harvesting." As the food source becomes more limited in the finishing phase of vermicomposting, described in Chapter 8, the worms will find less and less on which to feed and will naturally migrate to fresh food. So the first step toward harvesting is moving the castings to one side of the bin. Finished castings occur about five to ten days before the second step when you will actually harvest the castings for use.

When the finished vermicompost is ready to harvest, use a child's small leaf rake to scrape off the top two-inch (5cm) layer of vermicompost (castings). Wait for the worms, now visible, to retreat deeper into the vermicompost. Repeat this process until all the worms have been driven down into the bottom four or five inches (10-13cm) of castings.

Figure 47.
Harvest Method

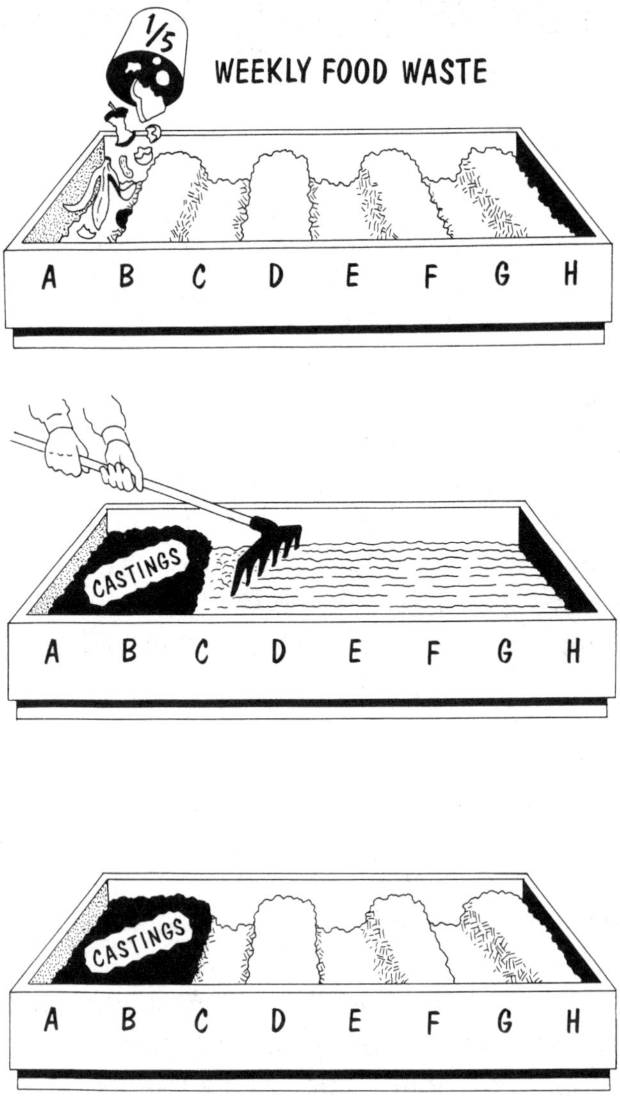

Next, shovel the remaining castings and worms at the bottom of the bin onto the bedding material on the right side of the bin (*see* Figure 47). Mix the bedding and turn the contents of the bin thoroughly. The earthworms that have migrated to the fresh food, combined with the worms driven to the bottom of the bin during top harvesting, will inoculate the bin with enough worms to begin again. The bin is now ready to return to regular use.

Each time castings are harvested some worms will also stay in the vermicompost. The worms harvested this way are good for the garden and will continue to do what they do best, improving the soil.

Figure 48. *Adding Compost to School Raised-Bed Gardens*

After Top Harvesting

The fresh vermicompost harvested may still have some small amounts of non-decomposed food waste and paper bedding in it. It can be used directly or be placed in a finishing box and held for later use (*see* Figure 39 for our location). During storage of the vermicompost, the worms harvested will continue to convert the waste which is left into castings.

To use the compost immediately, just screen it and use as a fertilizer for house plants or in gardens. Screening will remove unwanted food waste or paper.

At Laytonville we have extensive school gardens and we keep all of our castings in a finishing box when vermicompost is removed from the working bins. The finishing box (Figure 49) allows the vermicompost to continue breaking down,

Figure 49.
Finishing Box

allows all teachers and their classrooms access to the finished product, and protects it from the rain which would leach out valuable nutrients. Vermicompost is very popular in my district, and we never have an excess of castings.

Chapter 13

Project-Based Learning and Student Success

There are many of us in the classroom today who understand the importance of having students construct their own learning experiences. Education expert Robert Marzano's research shows us that the greatest learning gains occur when students add new experiences to what they have already learned and then construct new understanding from the added information. Students who apply newly learned information through meaningful work, experience the greatest learning gains. For learning to be powerful, you must engage the student in the planning stages as well as the process. At the same time, students need to make real connections to our world where complex social skills and problem-solving strategies are necessary for its future.

Students learn from activities which make them organize themselves in time. They learn from activities where they must keep appointments and schedule time to get things accomplished. Students need to explore rather than have content spoon fed to them. Teachers need to learn to become facilitators who guide their students in intellectual pursuits. Becoming the type of teacher who stands on the sidelines guiding students is not easy task. Most universities do not prepare their student teachers for this type of teaching. Teachers have become accustomed to a position as givers of wisdom.

Today, across the nation, students are often either disengaged or "acting out" in their classrooms. Students find themselves stuck in classrooms where the learning has little connection to any real or meaningful work. They often feel quite powerless and see little reason to do well in school. John Goodlad, in his book *A Place Called School*, discovered that not one single student among those interviewed in more than 2,000 hours of visitations to middle-school classes could coherently express a reason why they were in school. Students are not involved in any of the decisions about what they are to study, even though research indicates that students are more successful when they are involved and given choices

Figure 50. *Students Run Their Own Project*

about their own learning experiences. Schools which depart from the more traditional college-prep approach to learning, produce more successful students than schools which adhere to the current push for higher standards and the back-to-basics approach. Project-based learning activities soar beyond basics.

Our world is quickly becoming very complicated. Students need to develop skills which will help them become responsible, caring people who are capable of making good choices. If we do not offer them an opportunity to practice making choices, how are they to develop the skill? If we don't begin to help our students become critical thinkers, whose work is to be taken seriously, we are not truly preparing our students for life beyond school.

The more project-based learning you can conduct in your classroom, the more you will prepare your students for an ever-changing world, where the ability to communicate and cooperate with others and to solve complex global issues will help them get the job done. Ask yourself, "Do I need to do that? Can it be done just as effectively by a student?" How many opportunities do your students have to conduct research, plan activities, search for resources, or use technologies (telephones, fax machines, computers, *etc.*)? Ask yourself how much time you allow for conversations in your classroom. How much time do you allow for democracy in your classroom? When students are given these opportunities, their learning introduces them to life beyond the classroom and connects them with their communities and possible careers.

The greatest, most powerful, learning takes place when a student and a teacher learn together. Ask your students, "What do you want to learn?" The wonderful thing about worm science is that it is open-ended. Each and every bin you set up and study is a different world, with its own set of challenges. There is not one given formula that will work for each bin. Your students will need to use higher-level thinking skills to solve the problems they will encounter. Learning is not exciting when what is to learned is predetermined and the magic of exploration is lost.

I have included resources (*see* Appendix G) for you to use with your students. I encourage you to involve them in every step of the project, letting them add their own classroom culture to the project, to enrich their experiences and learning. Remember, there is much to be learned in the planning stages of a project. By having students do the planning and organizing themselves, their learning will be more meaningful and they will develop a greater understanding of the project. Our project was started because members of our student council wanted to make a difference in their community.

The following is a letter I received from the student of a neighboring school. His teacher had visited our school, and we had given her worms to carry back to her class. He was a student who didn't particularly like school, especially reading and writing. But something about worm science inspired him to painstakingly write the neatest letter he had written all year, reproduced here exactly as he wrote it:

> Dear Binet
> Thank you for getting the worms and saving the worms and feeding them well and helping the worms by building the worm bed and the worms need to protect the worms. Then the worms will say, "Thank you Mrs. Binet you are the worm's hero." You gather them up and make a worm bed and feed them. Then you give them rotten food that you know they will like
> Sincerely
> Jeff Patterson

I have seen many students motivated academically and socially by worm science. Because of its high level of engagement, many students who reluctantly do reading and writing projects, eagerly become great scientists, collecting and recording important information they feel has the potential to help the Earth's environment.

Project-Based Learning Checklist

Elizabeth Share and Laurette Rogers of The Autodesk Foundation in San Rafael, California, asked a group of teachers and administrators from 17 schools for a definition of project-based learning. They agreed that project-based learning needs to have the following characteristics to be a success:

- Teachers and administrators provide the framework for student decisions.

- There is a problem without a predetermined solution.

- Students design the process for reaching a solution.

- Students are in charge of accessing and managing the information they gather.
- Evaluation takes place continuously.
- Students reflect on what they are doing. What worked? What didn't work?
- A final product or goal results, and students evaluate it for quality.
- The classroom environment has an atmosphere which supports the importance of process.
- Students are encouraged to learn from their errors.
- Change is tolerated.

Accountability and Responsibility

Teachers are accountable by contract to the district that hires them, but they are responsible to their students and on-site colleagues, teachers elsewhere, their communities, and a vision of a better world. Accountability is a management term that employees, by signing a contract, owe their employers. Responsibility, on the other hand, is a personal commitment without legal obligation.

The responsibility to prepare children for a democratic world in which all participate to the best of their abilities is one teachers accept. Developing critical thinking is part of this preparation, and student management of projects reinforces this outcome. Sharing information with other teachers about their successes and failures is a minimal contribution a teacher can make to satisfy his or her personal responsibility.

Community businesses have the same contract/accountability relationship with their employees as school districts have with teachers. Meaningful participation by employees in the conduct of everyday business contributes to bottom-line financial success. The characteristics of project-based learning described above in school situations also apply to businesses and other organizations. Just substitute the terms "employees" for students and "employers" for administrators and teachers.

Using earthworms to compost food waste has been accomplished not only by homes, schools, and businesses, but also by nonprofit institutions. Where there is a will, there is a way!

At the Conservation Law Foundation in Boston, an audit showed that food waste accounted for about eight percent of the weight of their total solid waste. The Foundation started their vermicomposting with five pounds (2kg) of redworms and three Tupperware bins. Plastic collection containers for food waste were placed at strategic locations, for example, "coffee stations," *etc.* Comptroller Margaret Benson reports that using redworms to compost about 337 lbs (153kg) of food waste a year, or an average of 24 lbs (11kg) per person per year, avoids the $2.75 per bag cost the Foundation would otherwise have paid for waste pick up and disposal. Thus, vermicomposting food waste helped them cut the cost of waste disposal at their site.

This mid-scale vermicomposting manual with its appended materials is one way I can satisfy my responsibility to both students and other teachers. Use it productively in school and/or community settings!

Appendix A.

Checklist for Successful Mid-Scale Vermicomposting

1. Conduct a waste audit — this can be done by parents, community members or a group of students.

2. Get funding — Pacific Gas and Electric gives small mini-grants for projects that reduce waste generated on site. The American Association of Nurserymen gives grants to schools. The National Gardening Association in Burlington, Vermont, also gives small grants to schools.

3. Involve all staff from the very beginning of the project. Start with students and staff meeting together to develop a plan. Keep one another informed every step of the way.

4. Purchase the materials and have students assemble bins with staff or volunteers.

5. Start small. Start where you can and build up the program as you go. Don't wait.

6. Set up a schedule for waste flow. Provide clipboards for each bin.

7. Monitor and record the daily health of the bins.

8. Have students report their project to the school board. Invite local news personnel to the school. Have students conduct workshops on vermicomposting for the community.

9. Use the castings to grow vegetables in the school garden or for use to beautify the school grounds.

10. Know you are making a difference one step at a time.

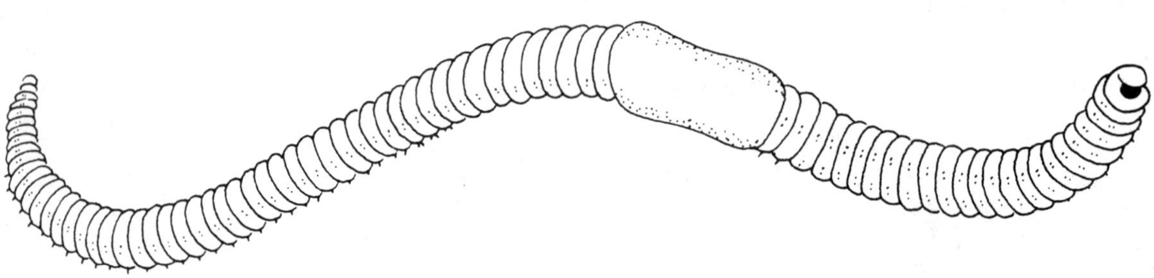

Appendix B.

Business Directory

Alpha Analytical Labs
860 Waugh Lane
Highway 1
Ukiah, CA 95482
(707)-468-0401

American Association of Nurserymen
Suite 500
1250 Eye Street NW
Washington, DC 20005
(202)-789-2900

American Community Gardening Association
100 North 20th Street, 5th Floor
Philadelphia, PA 19103-1495
(215)-988-8785

Autodesk Foundation
111 McInnis Parkway
San Rafael, CA 94903
(415)-507-5000

Beckley-Cardy Co.
1326 Ross Street, #B
Petaluma, CA 94954-6545
(707)-763-9900
(800)-446-1477

Center for Ecoliteracy
2522 San Pablo Avenue
Berkeley, CA 94702

Conservation Law Foundation
62 Summer Street
Boston, MA 02110-1016
(617)-350-0990

Food Works
64 Main Street
Montpelier, VT 05602
(802)-223-1515

Harwood Products
1 Main Brascomb Road
Brascomb, CA 95417
(707)-984-6181

Interactive Learning Systems
214 Water Street
Point Richmond, CA 94801

National Gardening Association
180 Flynn Avenue
Burlington, VT 05401
(802)-863-1308

North Coast Rural Challange Network
Mendocino Unified School District
P.O. Box 1154
Mendocino, CA 95460
(707)-937-5868

continued next page

Pacific Gas & Electric Co. (PG&E)
Office of the Corporate Secretary
One Market, Spear Tower, Suite 2400
San Francisco, CA 94105
(415)-267-7070

Rural Challenge
P.O. Box 1569
Granby, CO 80446

Appendix C.

Posters

Today we know that children learn in a variety of ways. I always post the following information in my classroom so that parents and children understand the importance of experiential learning. There is no more powerful tool for learning than involvement in a task, not only intellectually, but physically. "Hands on" learning prepares the learner to teach others most effectively.

Dots appear on posters that are larger than one page. Photocopy poster pages for your own use. Align the dots on adjacent copied pages and cut off excess margin on dotted line to assemble a given poster.

Learning and Action

Hear, See, and Do

All *Real*

Learning

Is

Connected

to

Action

--I HEAR AND I FORGET

--I SEE AND I REMEMBER

--I DO AND I UNDERSTAND

Chinese Proverb

Appendix D.

Forms

Dear Parents

Cafeteria Waste Audit

Paper Waste Audit

Places to Look for Waste

Daily Record Sheet

Date____________________

Dear Parents

Our class will be involved in variety of hands-on activities using earthworms. We will be learning a variety of scientific and mathematical skills as we look at the importance of earthworms to the environment. These life science activities will give us an up-close and personal look at ecosystems and ecology. Ecology is the study of how nature works together as a whole. Ecology is an exciting subject because it is, to a certain extent, the last wild frontier. We will be studying the ecology of a worm bin and the plants and animals we find there while they are living, breeding, being consumed, growing, dying, and eventually decomposing. As energy flows through a fascinating worm bin ecosystem, it is a "dog-eat-dog" world in there. As scientists, we will be studying the complex relationships we find in the world of a worm bin.

As human beings we are discovering that our relationship with the environment must change if we are to survive as a species. Our short-sighted activities have left us with limited resources. Pollution of the environment is a natural occurrence. In a healthy ecosystem these wastes are recycled and a balance is maintained. If students can develop a greater understanding of their relationships by studying the complex relationships found in a worm bin, maybe we can use natural relationships as a guide to keep our environment safe and stop the pollution of our water, air, and soil. During this unit of study, students will be bringing home activities to share with you. I encourage you to learn along with your child. There is no more powerful tool for learning than being responsible for teaching the concept to someone else.

Our study will include the study of fungi. If your child has allergies to mold, mildew, or penicillin, for example, please indicate in the space provided below. Students will be recycling school food waste through a process of vermicomposting. If you have any questions about this unit of study, please call me at ________________________. I always look forward to teaching this unit and invite you to visit us here at school.

Sincerely,

TEAR OFF AND RETURN TO TEACHER IF RELEVANT: My child ________________________ has an allergy to molds. Please call me at ________________ for further information.

Cafeteria Waste Audit

Week Number______ **Date_______________**

1. The total number of lunches served per day: _____________________

2. Daily lunch veggie waste in weight: _______ lbs and __________ounces to be vermicomposted.

3. Daily lunch meat and protein waste in weight:_______lbs and_______ounces.

4. Types of uneaten food thrown away:

____

____

____

5. Foods that are consistently tossed into the compost:

____

6. Dumpsters: Size___________ Number_____ Collection Frequency___________

7. Are milk cartons being recycled? ________

8. Are bottles being recycled? ________

9. Are aluminum cans being recycled? ________

10. Are plastics being recycled? ________

11. Are paper bags being recycled? ________

12. Are janitors and personnel involved in recycling? ________

13. Is tin being recycled? ________

14. Thoughts on current recycling efforts at our school: __________________________

____

Paper Waste Audit

Week Number______ **Date_______________**

Track the paper waste in your business or school. Don't forget to include the office in your paper audit. Find out the type and quantity of paper being discarded. The non-styrofoam packing used in boxes is made from starch and can be safely put into worm bins.

Paper types	***Quantity***	***Paper types***	***Quantity***
( ) Copy paper	_______	( ) Newspaper	_______
( ) Colored construction paper	_______	( ) Magazines or book orders	_______
( ) Primary paper	_______	( ) Cardboard	_______
( ) Note book paper	_______	( ) Mixed paper waste	_______

1. How many classrooms do you have? __________
2. How many classrooms are currently recycling their paper? __________
3. How many recycling containers will you need for each classroom? __________
4. Is your school or business currently recycling any of their paper waste? __________
5. Where does the paper waste from your offices go? __________________________

____

Places to Look for Waste

— Waiting rooms and entry areas

— Employee lunch rooms

— Copy rooms near copy machines

— Meeting rooms

— Lounge areas

— Employee work stations

— Classrooms

— Near paper shredders

— Kitchens

— Cafeterias

— Libraries

— Computer centers

— Resource rooms

— Board rooms

Daily Record Sheet

Use this record sheet to help you record and remember where you buried your food waste. Each bin should have its own clipboard and record sheet. Bins should be numbered.

Initial Set-Up Date__________

Today's Date__________

Outside Temperature________

Inside Temperature of Bin________

Bin Number_________

Food Weight________

Trench A B C D E F G H

Added Paper Yes No

Condition of Bin____

____

____

____

____

____

Appendix E.

Lesson Plan: Cross-Age

The plan on the next page is used by cross-age tutors and parent or community volunteers. When parents volunteer, sometimes it is difficult for them to conduct a mini-lesson. I developed this form to help them prepare for the lesson. I found that middle-school students were very successful with this plan.

Before students use this lesson with younger students, they should practice the lesson with one another. Lesson plans should always be checked for content before using with children to ensure accuracy.

Cross-Age Lesson Plan Guide

Subject Area/Activity Name:

Procedure:

Concept to Emphasize:

Questions to Ask to Check for Understanding:

Number of Tutors: _______ Number of Children: ______

Limits:

Strategies to Encourage Good Behavior and Stimulate Interest in the Activity:

Appendix F.

Work Sheets

Pre-Test/Post-Test

Observation

Using the Scientific Method

Worm Tender's Record Sheet

Vermicomposting Record Sheet

Earthworm Anatomy

The Muscles

Be a Vermicompost Explorer

Worm Castings — Picture Journal

Record of Plant Growth

Make a List

Worm Questions: *Worms Eat My Garbage,* Chapters 1-14

Earthworm Science — Quarter Final

Worms, True or False
Pre-Test / Post-Test

This should be filled out by all students before you begin a unit on earthworms. It can be given again later as a post-test. I often have the students take it home and give it to their parents once they know the answers themselves.

_______1. Worms have eyes.

_______2. Worms have teeth.

_______3. Worms have bristles on their bodies.

_______4. Worms have blood which is similar to ours.

_______5. Worms have both sexes.

_______6. If you cut a worm in half, both parts will grow back and you will get two worms.

_______7. Worms can live for four years.

_______8. Worms hatch from cocoons.

_______9. Worms love to swim in puddles when it rains.

_______10. Worms live underground and do not need air.

_______11. Worms breathe through their skin.

_______12. One pound of worms can eat one pound of food waste per week.

_______13. A worm's skin must be moist for them to breathe.

_______14. Worms like to eat paper because it has lots of vitamins they need for health.

_______15. Castings are little braces they wear on their legs.

_______16. Worms have gizzards like chickens.

_______17. Worms used for vermicomposting can also be used for fishing.

_______18. Worms are photosensitive and have light-sensitive cells.

_______19. Worms are important because they aerate the soil.

_______20. Worms are sensitive to touch.

Observation

This is a generic form which can be used for many different worm bin observations with young students. The type of observation would be listed on line 1. Explain to your students that detailed observations are very important to scientists because they use these observations to develop scientific facts.

Name_______________________________

1. ____________________________________Observation

2. Date and Time_____________________________

3. I looked at____

4. Below is a picture of what I saw:

5. Here are some things I noticed:

____

____

____

Using the Scientific Method

This form is used with primary students, either individually or in groups, to introduce them to the scientific method. In this experiment they will observe the effects of worm castings on the germination and growth of plants. Each student should have four paper cups and a small amount of soil for the seeds. In two of the cups, use worm castings. In the other two cups, use any type of soil and no worm castings. Plant two seeds in each cup and water them. Any type of seed will do for this experiment. With small children I would use large seeds because they are easier to handle.

Name (s)________________________________

1. This experiment is called_________________________.

2. It was started on____________________.

3. The question is____

____.

4. The hypothesis is____

____.

5. To test the hypothesis, I did this experiment (list each step):

____

____

____

____

6. Seed sprouts were first seen on____________________

7. Is the color of each plant the same?__________

Is one group of plants greener than the other?____________

Why do you think one plant might be greener than another?____________

____

8. Which seeds sprouted first?______________________________________

____

9. Which seeds grew fastest?______________________________________

10. Which seeds grew tallest?______________________________________

11. Please add one observation of your own__________________________

____

12. My conclusion is____

____

____

____

____

Worm Tender's Record Sheet

This sheet can be used successfully with primary students. Break them up into small groups and give each student a job to perform so everyone participates.

Names:

Worm Tender____________________________

Facilitator ___________________________

Recorder ___________________________

Materials Manager___________________________

Researcher(s)___________________________

Reporter___________________________

Harmonizer___________________________

Date____________________

Time____________________

Temperatures:

Outside Air __________

Inside Air __________

Inside Worm Bin __________

1. Look for worms moving away into the dark part of the bin, insects and other animals, food, bedding, mold, and worm castings.

- Is the food breaking down?__________
- Record what you see in the worm bin:

2. Stir up the contents **carefully** from top to bottom.
   - Do you see any egg cases or baby worms? __________
   - Record what you see now:

3. Remember your numbers must add up to 100 percent and estimate what percentage of the worm bin is:

food __________

bedding __________

worms __________

castings __________

100%

4. Add if necessary:

water Is everything moist enough? _______ How can you tell?

soil Is there grit for digestion? ________

food Is there too much or too little food? __________________

bedding Can the worms escape from their castings?_________

Vermicomposting Record Sheet
Intermediate Level

Name ______________________________ Date_____________________

Description of Set-up:

Breeders______ Mixed Sizes________ Cocoons________

Type of Bedding Used________________

Size of Bin________________________ Surface Area____________________

Food Waste Buried: Weight_______ Variety__________________________

1. Based on bin surface area and weight of worms, how much food waste should be fed

Per day?_____________ Per week?_____________ Per month?_____________

Observations:

2. Does the bedding seem to be dry around the edges? ___________ Is the bedding moist enough for the worms?__________ How can you tell?

3. What is the outside temperature?________

   What is the internal temperature of the bedding?________

4. Is there an anaerobic or an aerobic odor present in the bin? ________
   How can you tell?

5. Where are the worms located? ______________________

   Are they congregating in one particular area? ________
   If the worms seem to be in one area, why do you think this is happening?

6. Are the worms feeding around something that especially appeals to them? ________
   If so, record what that something is and why you think they are there.

7. Are any worms mating? ________

8. Are there any cocoons present in the bedding?________
   If so, how many did you find?

9. Are the cocoons located on top, or throughout the bedding? ______________________

10. Are there any baby worms present? ________

11. Are there any insects present in the bedding? ________
    If so what are they and how did they get there?

12. Are there castings present in the bedding? ________

13. Estimate what percentage of the bin is:

    food ____________

    bedding ____________

    worms ____________

    castings ____________

    cocoons ____________

    100%

14. What needs to be added?

15. Please add an observation of your own about your vermicomposting system.

Earthworm Anatomy

Use these prompts with a set of anatomy drawings:

1. How many hearts or aortic arches do you find? ____________________
2. What organ functions as a brain? ____________________
3. What organ helps an earthworm grind up its food? ____________________
4. This organ allows the earthworm to breathe. ____________________
5. This organ produces the eggs. ____________________
6. This organ removes waste. ____________________
7. This carries blood to the skin and to the intestine. ____________________
8. This vessel carries blood to the aortic arches. ____________________
9. These organs receive sperm. ____________________
10. These vessels store sperm. ____________________
11. This organ stores food. ____________________
12. This organ swallows the food. ____________________
13. Food passes through what after being swallowed?____________________
14. What digests food in the worm's body? ____________________
15. What will you find in the gizzard? ____________________

The Muscles

1. An earthworm's muscles cause it to move by contracting and relaxing. There are two layers of muscle tissue:

   One layer is called the ________________________and circles the worm.

   The other layer lies lengthwise and is called________________________.

2. Draw a cross-section of an earthworm. Label the muscles.

Be a Vermicompost Explorer

Take an excursion inside a worm bin and discover for yourself what lives in the waste frontier. You may need to use a magnifying glass to see some of the smaller organisms living in the bin.

Can you find any of the following? Draw one or two of them.

Sow bug _______

Centipede _______

Seed _______

Cocoon _______

Adult redworm _______

Baby worm _______

Mold _______

Mushroom _______

Frog _______

Newt _______

Mite _______

Fungi _______

Slug _______

Snail _______

Millipede _______

Worm Castings — Picture Journal

Name________________________________ Date____________

BOX 1. Plant with worm castings.	BOX 2. Plant with no castings.

Record of Plant* Growth

Start Date of Project:	Height of Plants in Centimeters	
	With Castings	Without Castings
Day		
Day		
Day		
Day		
Day		
Day		
Day		
Day		
Day		
Day		

*PLANT NAME(S) ________________________________

Make a List

1. List the things you throw away each day at home.

2. List the things you throw away at school.

3. Compare the two lists (1 and 2 above). Brainstorm ways you can improve your trash habits.

4. What could you do with some of the items you throw away?

5. What organic things do you throw away? What could you do with them instead of throwing them away?

Worm Questions

All of the following questions were developed to be used with Chapters 1-14 of Mary Appelhof's book, *Worms Eat My Garbage* (WEMG). When I teach worm science at the middle-school level, I have students use her book as a textbook.

Worms Eat My Garbage, Chapter 1:

1. The process of converting food waste into compost using red worms is called

____

2. A worm bin system consists of five parts. Name and describe each part.

____

____

____

____

____

3. Many containers make acceptable worm bins. List two.

____

____

Worms Eat My Garbage, Chapter 2:

1. Your redworms need certain requirements to be content. At what temperature should you keep them? ____________

2. Why do redworms need moisture? ____________
What will happen if your worm bin is flooded with water? ____________

____

3. Redworms can tolerate a wide range of acidity in their environment. What is the correct pH range for a worm bin? ____________

4. Redworms need oxygen to breathe and they produce carbon dioxide. What would happen if you wrapped a worm bin in plastic and allowed no ventilation? ____________

____

5. Vermicompost is composed of more than just worm castings. What might you find in vermicompost besides worm castings? ____________

____

6. What might happen to the redworms in your bin if the bin completely converts to vermicast? ____________

7. Compare and contrast vermicomposting systems at high, medium, and low maintenance levels.

8. How do vermicomposting systems differ in maintenance between mid-scale and home set ups? ____________

____

9. Where can worm bins can be located? ____________

10. What three things must you consider when deciding on a location for your worm bin?

____________ ____________ ____________

Worms Eat My Garbage, Chapter 3:

1. When looking at containers to use for your worm bin, why is the surface area of the container important? ______________________________

____

2. What is the depth of the ideal worm bin? ____________

3. Define aerobic. ______________________________

____

4. Define anaerobic. ______________________________

____

5. How do you decide how large your system should be? ______________

____

____

6. What is the ratio of food waste to surface area? ______________

7. Why is it unwise to use pressure-treated lumber? ______________

____

____

8. What are the advantages to using wood rather than plastic? ______________

____

____

9. List two reasons for drilling holes into the worm bins.

____________________ ____________________

10. There are many variations of worm bins. Pick a worm bin that you think will work the best and tell me why you think it is superior to other bins.

Worms Eat My Garbage, Chapter 4:

1. You can use many types of beddings. List them, as described in Chapter 4, and give an advantage and disadvantage of each.

______________ ______________________ __________________

______________ ______________________ __________________

______________ ______________________ __________________

______________ ______________________ __________________

2. Why should we not use peat moss as bedding material?______________

____

3. What are two reasons to add soil to the bedding?

__________________________ __________________________

4. Why would you use calcium carbonate?______________________

5. What could you use instead of calcium carbonate?______________

6. What does rock dust contribute to the vermicompost?______________

____

How does rock dust affect plant growth? ______________________

7. What is Zeolite?______________________________________

8. How is Zeolite used?___________________________________

____

9. What country manufactures Zeolite?________________________

10. Define a "green business."______________________________

____

____

____

Worms Eat My Garbage, Chapter 5:

1. Most of the worms we dig up while gardening are not vermicomposting worms. List three reasons why you cannot raise garden worms.

___

___

___

2. What are some distinguishing features of the red worms we use for vermicomposting?

___ ___ ___

3. Scientists have a system for naming organisms using just two words. The first word is called the genus and the second word is the species name. Both of the words come from what two languages other than English?

___ ___

4. What is the full scientific name for the redworms that Mary Appelhof uses in her vermi-composting bin? ___

5. Are the *E. fetida* used in vermicomposting equipped for living in the ground?___

6. What worm is also called a dew worm or angle worm? ___

7. What happened when Mary put nightcrawlers in her worm bin? ___

___

8. What important part do nightcrawlers play in soil fertility? ___

___

___

9. Is it easy to identify redworm species?___Why is this so?___

___

10. Why is it important to use a scientific name when buying your worms?___

___

Worms Eat My Garbage, Chapter 6:

1. Define clitellum. ______________________________

____

2. Worms hatch from egg cases called ______________.

3. How many eggs do you find in an egg case? ______________

4. How many baby worms will emerge from the egg case? ______________

5. What are three conditions that affect how many hatchlings will emerge?

____

____

____

6. What are the factors that affect when a hatchling becomes a breeder? ______________

____

____

7. A worm can be sexually mature in how many weeks? ______________

8. When sexually mature, how many cocoons can it produce per week? ______________

9. What controls the size of a worm population? ______________________________

____

____

____

10. What will happen if you stabilize the amount of food waste you give to your worm bin population? ______________________________

____

Worms Eat My Garbage, Chapter 7:

1. We measure a quantity of worms by their ___________ rather than their numbers.

2. What is meant by a biomass?____

____

____

3. What are the three factors which will influence how many worms you start with?

____

____

____

4. What is the worm to food-waste ratio?__________

5. What is the difference between breeders and bedrun?

____

____

Roughly, how many hatchlings are in a pound of bedrun worms?___________________

Worms Eat My Garbage, Chapter 8:

1. To prepare bedding for a worm bin, what five items do you need?

____________________ ____________________ ____________________

____________________ ____________________

2. The amount of bedding your bin needs depends on the _____________ of your container.

3. You should use roughly about _________ pounds of bedding per cubic foot of the bin.

4. What is the water to bedding ratio? ______________________

5. What are the steps when setting up a worm bin?_______________________________

____

____

____

____

____

6. What percent of moisture should the bedding have?____________________

7. Because water can accumulate in plastic bins, you should reduce the water added by what fraction? _________________

8. Why should you add water to manure and let it set for 2 or 3 days before you add the worms? ____

9. How can you tell if bedding is too wet?______________________________________

10. Why do growers package and mail their worms in drier bedding?_________________

____

Worms Eat My Garbage, Chapter 9:

1. What toxin can you find in citrus? ____________________

2. Four reasons we do not add meat to a worm bin:

________________________________ ________________________________

________________________________ ________________________________

3. Can worms eat meat? ____________

4. In India they use worms to process what type of waste? ____________________

5. Meat is a highly ______________ -rich material.

6. Name five things not to put in your worm bin, and tell why not for each.

__________ ____

__________ ____

__________ ____

__________ ____

__________ ____

7. Why is it important to keep animal feces ***out*** of your worm bin? ____________________

____

8. It is important to bury the food waste in the bedding because ____________________

____

9. Why do we put a plastic cover over the contents of the worm bin? ____________________

____

10. What happens to a worm bin when you overload the system? ____________________
    How can you help correct an overloaded system? ____________________

Worms Eat My Garbage, Chapter 10:

1. When adding food waste to bedding, it is important to make certain observations. What are four things you should look for?

______________________ ______________________

______________________ ______________________

2. Why is it important to keep records? ______________________

3. Why does moisture collect in plastic worm bins? ______________________

4. What can you do with excess water run-off from a worm bin? ______________________

____

____

5. What does adding dry bedding do to excess moisture in bins? ______________________

6. Two kitchen tools used for draining off excess moisture:

______________________ ______________________

7. What is a peat moss worm? ______________________

____

8. Peat moss is used for ______________________

____

9. How many weeks will pass before you see a noticeable difference in your worm bin's bedding? __________

10. Describe two methods of worm harvesting. ______________________

____

____

Worms Eat My Garbage, Chapter 11:

1. Does a worm have eyes? ___________ What makes worms sensitive to light?

____

2. Do worms have teeth? ___________ How do they digest their food? _______________

____

____

3. A prostomium is ____

4. Where does food go once it leaves the gizzard? ____________________________

____

5. How do worms get air underground? ______________________________________

____

6. If you cut a worm in half, will two worms grow from the severed parts? ___________

7. What happens to worms that die in the bin? ______________________________

____

8. How long can redworms live? ___________

9. How long do *most* redworms live? ___________

10. What should you look for if many worms are dying in your bin? ______________

____

____

Then what should you do? ____

Worms Eat My Garbage, Chapter 12:

1. In a worm bin energy flows from organism to organism. In time, after you build your worm bin, other organisms appear. What are first-level consumers? ______________

____

What are second-level consumers? ______________________________

What are third-level consumers? ______________________________

2. What do enchytraeids eat? ______________________________
3. What are springtails? ______________________________
4. Isopods are related to what other animals? ______________________________

   How do they breathe? ______________________________
5. How can centipedes be dangerous in your worm bin? ______________________
6. Northern Ireland has a problem with ____________ that have reduced their earthworm populations.
7. What type of mite is a problem in worm bins, and how can you get rid of them?

____

____

8. How could you attract flies to your worm bin? ______________________________

   If they become a problem, how can you get rid of them? ______________________

____

9. If you have a problem with ants, what can you do to cure the problem? ______________

____

10. If you are allergic to molds, should you have a worm bin in your home? __________

    Why or Why not? ______________________________

____

Worms Eat My Garbage, Chapter 13:

1. A worm casting is a ________________________________ and contains thousands of ____________________.
2. What is humus? ____

   Why is it important to soil and plant growth? ____________________________

   ____
3. Using vermicompost with seedlings will not ________ your plants like many fertilizers do.
4. In what ways do young plants sprouted from seeds benefit from vermicompost?

   ____

   ____
5. Why is vermicompost good for young transplants? _________________________

   ____
6. How can vermicompost be used as a top dressing?_________________________

   ____
7. Describe vermicompost after several months of low maintenance vermicomposting.

   ____

   ____
8. What happens if you let the vermicompost dry out?__________________________

   ____
9. What are the advantages for not sterilizing vermicompost? ___________________
10. List the ingredients of a good potting mix. ____________________________

    ____________________________ ____________________________

    ____________________________ ____________________________

Worms Eat My Garbage, Chapter 14:

1. What are some of the advantages of using a worm bin to convert your food waste to vermicompost? ____

____

2. What are some of the benefits of recycling? ____________________________

3. What are the three Rs, and what does each of them mean?

________ ____________________________________

____

________ ____________________________________

____

________ ____________________________________

____

4. What is post-consumer paper?____________________________________

____

5. How do you benefit financially from recycling?____________________________

____

____

How does the planet benefit?____________________________________

____

____

Earthworm Science — Quarter Final

1. How many hearts does an earthworm have, and what is the scientific name of these organs?

   ______ ______________________________

2. What does anaerobic mean? ______________________________

   ____

3. What does aerobic mean? ______________________________

   ____

4. How does an adult earthworm differ from a juvenile earthworm? __________

   ____

   ____

5. An earthworm has both male and female sex organs. What is this condition called?

   ____

6. An earthworm hatches from a ______________________________.

7. Of the many things an earthworm does for the environment, list two of them:

   ______________________ ______________________

8. Temperatures between__________and__________are safe for earthworms.

9. Since earthworms have no teeth, how do they "chew" their food?

____

____

____

10. What other animal is the main predator of an earthworm? __________

11. ______________________ means composting with redworms.

12. When does an earthworm become sexually mature? ____________

____

13. Earthworms crawl along the ground when it rains. Why do they do this?

____

____

14. Name two types of beddings you can use in worm boxes:

____________________ ____________________

15. Explain how we use earthworms to compost our food waste. Please include every step:

____

____

____

____

____

16. Turn over these test papers. Draw an earthworm's body both external and internal on the blank sheets. Include as many parts as you can remember.

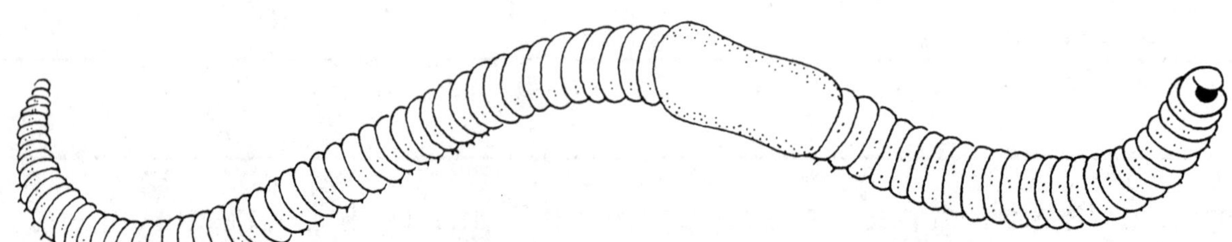

Appendix G.

Annotated Classroom Materials

Each of the two major sections in this appendix is divided in four subject catagories as follows:

Publishers' series appear in parentheses after the date of the bibliographic descriptions. Appropriate grade levels are suggested at the right margin of the line following the book annotation.

I. Children's Books

ANIMALS

AMAZING SPIDERS, Alexandra Parsons. Photographs by Jerry Young. New York, NY: Alfred A. Knopf, 1990. (Eyewitness Books)

A close-up look at some of the worlds most amazing spiders.

Grades K-3

AMAZING WORLD OF BUTTERFLIES AND MOTHS, Louis Sabin. Illustrated by Jean Helmer. Mahwah, NJ: Troll Communications, 1982.

This book compares and contrasts moths and butterflies.

Grades K-2

BACKYARD BUGS, Joanne Nelson. Illustrated by Roberta Collier. Columbus, OH: Modern Curriculum Press, 1989.

This book is a predictable science story. Easy reader.

Grades K-1

<u>BACKYARD INSECTS</u>, Millicent E. Selsam. Photographs by Ronald Goor. New York, NY: Scholastic, 1988.

This book tells you how to find more than 1,000 different insects that may live in your backyard.

Grades K-2

<u>BEES</u>, Gallimard Jeunesse. Illustrated by Ute Fuhr and Raoul Sautai. New York, NY: Scholastic, 1992.

Young children can meet a world of bees in this beautifully detailed book.

All Ages

<u>THE BIGGEST HOUSE IN THE WORLD</u>, Leo Lionni. New York, NY: Alfred A. Knopf, 1987.

In a fable within a fable, a wise snail convinces his son that a small, easy-to-carry shell might be better than the biggest house in the world for a life of adventure and exploration.

Grades K-1

<u>THE BIRD ALPHABET BOOK</u>, Jerry Pallotta. Illustrated by Edgar Stewart. Watertown, MA: Charlesbridge Publishing, 1989.

The author, who has written similar books about bugs, flowers, the ocean, and trash, uses the alphabet to explore each topic.

Grades K-2

<u>BUGS</u>, Joan Wright. Illustrations by Nancy Parker. New York, NY: Scholastic, 1988.

Rhyme is combined with detailed descriptions of 19 bugs.

Grades K-2

<u>BUTTERFLIES AND MOTHS</u>, George S. Fichter. Illustrated by Kristin Kest. New York, NY: Western Publishing, 1993. (Golden Junior Guide to Butterfly Observations)

This book illustrates the differences between moths and butterflies.

All Ages

<u>BUTTERFLY</u>, Mary Ling. Photographed by Kim Taylor. New York, NY: Dorling Kindersley, 1992. (See How They Grow)

This book has dramatic photographs that chronicle the life cycle of a swallowtail butterfly. Beautiful!

Grades K-2

<u>BUTTERFLY</u>, Michael Chinery. Mahwah, NJ: Troll Communications, 1991.

Beautiful photographs trace the life-cycle of a swallowtail butterfly.

Grades K-2

<u>THE BUTTERFLY ALPHABET</u>, Kjell B. Sandved. New York, NY: Scholastic, 1996.

Butterfly wings mirror the images of the alphabet. Beautiful book that can be used as a prompt for writing poetry.

All Ages

<u>BUTTERFLY AND CATERPILLAR</u>, Barrie Watts. Morristown, NJ: Silver Burdett, 1985.

Watch a caterpillar turn into a butterfly in the remarkable photographs found in this book

Grades 1-3

<u>THE BUTTERFLY HUNT</u>, Yoshi. Saxonville, MA: Picture Book

Studio, 1991.

Yoshi's wonderful expressive pictures capture the important moment when other creatures on the earth become real to a child, and suddenly have the right to live their own lives.

Grades K-2

CHARLIE THE CATERPILLAR, Dom DeLuise. Illustrated by Christopher Santoro. New Bern, NC: Golden Book Club, 1993.

A story of friendship and metamorphosis.

Grades K-1

DIG HOLE, SOFT MOLE, Carolyn Lesser. Illustrated by Laura Regan. New York, NY: Harcourt Brace, 1996.

Enter the world of the star-nosed mole. This is a beautiful book about a little known mole who can live both in the soil and in the water as well.

Grades K-2

EARTHWORMS, Chris Henwood and Barrie Watts. London, UK: Franklin Watts, 1988. (Keeping Small Mini-beasts)

This has wonderful photographs of earthworms. It is one in a series of "Keeping Small Mini-beasts," that also includes Ants, Beetles, Butterflies and Moths, Caterpillars, Frogs, Grasshoppers and Crickets, Ladybugs, Snails and Slugs, and Spiders.

Grades K-4

FEATHERS FOR LUNCH, Lois Ehlert. New York, NY: Harcourt Brace Javanovich, 1990.

Beautiful illustrations.

Grades K-2

THE FRIENDSHIP OF HESPER AND RANI, Letitia U. Soloman. Illustrated by Marsha Mello. Calpella, CA: Henchanted Books, 1996.

This book explores the friendship of a caterpillar and a tadpole, showing metamorphosis.

Grades 1-2

THE GROUCHY LADYBUG, Eric Carle. New York, NY: Scholastic, 1977.

Carle's book about a grouchy ladybug who flies through the hours of the day.

Grades K-2

THE HUNGRY CATERPILLAR, Eric Carle. Saxonville, MA: Picture Book Studio, 1987.

The story of a tiny, very hungry caterpillar who eats his way through the story, counting.

Grades K-1

THE ICKY BUG ALPHABET BOOK, Jerry Pallotta. Illustrated by Ralph Masiello. New York, NY: The Trumpet Club, 1989.

This is a fun book. The author, who has written similar books about birds, flowers, the ocean, and trash, uses the alphabet to explore each topic.

Grades K-2

INSECTS DO THE STRANGEST THINGS, Leonora and Arthur Hornblow. Illustrations by Dorthea Barlowe. New York, NY: Random House, 1990. (Step-Up Nature Books)

This book explores the life cycle and habits of a variety of insects.

Grades K-2

JOYFUL NOISE, POEMS FOR TWO VOICES, Paul Fleischman. Illustrated by Eric Beddows. New York, NY: Harper and Row, 1990.

Poems about insects.

Grades K-2

THE LADYBUG AND OTHER INSECTS, Gallimard Jeunesse and Pascale Bourgoing. Illustrated by Sylvie Perols. New York, NY: Scholastic, 1989.

Beautiful overlays illustrate the life cycle of a ladybug.

Grades K-2

LUNCH, Denise Fleming. New York, NY: Scholastic, 1996.

Crunch! Munch! Is mouse eating again? A very hungry mouse eats his way through many fruits and vegetables.

Grades K-2

MAMMALS, HOW TO WATCH AND UNDERSTAND THE CAPTIVATING WORLD OF MAMMALS, David Burnie. New York, NY: Dorling Kindersley, 1993. (Eyewitness Books)

All the Eyewitness Books present information in a very visual way.

All Ages

MISS SPIDER'S TEA PARTY, David Kirk. Hong Kong, BCC: Palace Press, 1995. (Callaway Editions)

This is one book in a series of Miss Spider stories which captivate you with rhyme and bold color. It can also be used as a prompt for writing poetry at all grade levels and the brillant drawings can be used in art classes. My granddaughters love this series. Other books in this series include *Miss Spider's New Car* and *Miss Spider's Wedding*.

Grades K-2

MONARCHS, Kathryn Lasky. Photographs by Christopher G. Knight. New York, NY: Harcourt Brace, 1993.

This beautiful book traces the journey of the monarch and the people whose lives they touch in very special ways.

Grades 4-8

A NOT SO UGLY FRIEND, Stan Applebaum and Victoria Cox. Illustrated by Tony Chen. New York, NY: Holt, Rinehart and Winston, 1983.

This book explores the characteristics of an earthworm. Beautiful illustrations!

Grades K-1

A PICTURE BOOK OF BUTTERFLIES AND MOTHS, Joanne Mattren. Illustrated by Roseanna Pistolesi. Mahwah, NJ: Troll Communications, 1996.

Beautiful illustrations of butterflies and moths.

Grades K-2

THE SNAIL'S SPELL, Joanne Ryder. Illustrations by Lynne Cherry. New York, NY: Scholastic, 1982.

Outstanding science book for young children. "Imagine you are soft and have no bones." Students explore the universe of a snail.

Grades K-1

SQUIRMY WORMY COMPOSTERS, Bobbie Kalman and Janine Schaub.

New York, NY: Crabtree Publishing Company, 1992. (Primary Ecology Series)

A book which details an inside vermicomposting unit. Great pictures of students working with their worms. Great book!

Grades K-4

STELLALUNA, Janell Cannon. New York, NY: Scholastic, 1993.

A beautiful story about the adventures of a little fruit bat.

Grades K-3

THE ULTIMATE BUG BOOK, Luise Woelflien. Illustrated by Wendy Smith-Griswold. New York, NY: Western Publishing, 1993.

This book explores the fascinating life of bugs.

Grades K-5

WHAT DO YOU WANT TO KNOW ABOUT EARTHWORMS, Seymour Simon. New York, NY: Scholastic, 1969.

This book gives you information about earthworms.

Grades 4-6

WHAT'S INSIDE INSECTS? Angela Royston. Illustrated by Richard Manning. New York, NY: Dorling Kindersley, 1992. (Scholastic Book Club Edition)

Where are a ladybug's wings? How does a fly walk upside down? These questions and more are answered in this beautiful book.

Grades K-2

WHAT'S INSIDE SMALL ANIMALS, Angela Royston. Illustrated by Richard Manning. New York, NY: Scholastic, 1995.

This book is designed to help young children understand what goes on inside the bodies of small animals.

Grades K-2

WHERE BUTTERFLIES GROW, Joanne Ryder. Pictures by Lynne Cherry. Hydesville, CA: Lodestar Books, 1989.

An imaginative journey that transforms a caterpillar to a black swallowtail butterfly. You can also enjoy discovering how to grow your own butterfly garden.

Grades K-2

WONDERFUL WORMS, Linda Glaser and Loretta Krupinski. Brookfield, CT: Millbrook Press, 1992.

This is a friendly book about earthworms and it is filled with facts.

Grades K-2

PLANTS

THE ACORN'S STORY, Valerie Greeley. New York, NY: Macmillan Publishing, 1990.

A book with beautiful illustrations which tells the story of an acorn.

Grade K-2

ANNA'S GARDEN SONGS, Mary Q. Steele. Pictures by Lena Anderson. New York, NY: Greenwillow Books, 1989.

Poetry and beautiful water colors help students explore the world of vegetables.

Grades K-2

ANNA'S SUMMER SONGS, Mary Q. Steele. Pictures by Lena Anderson. New York, NY: Scholastic, 1988.

Trees, flowers, ferns, and fruit—Anna's joyous songs celebrate everything that grows.

Grades K-2

APPLES AND PUMPKINS, Anne Rockwell. New York, NY: Scholastic, 1989.

When red and yellow leaves are on the trees, it's time to pick apples and pumpkins.

Grades K-2

THE BIGGEST PUMPKIN EVER, Steven Kroll. Illustrated by Jeni Bassett. New York, NY: Scholastic, 1984.

Two delightful mice fall in love with the same pumpkin. Cute!

Grades K-2

A BUSY YEAR, Leo Lionni. New York, NY: Scholastic, 1992.

Willie and Winnie love their tree all year long.

Grades K-2

CALIFORNIA WILDFLOWERS, A CHILDREN'S FIELD GUIDE TO THE STATE'S MOST COMMON FLOWERS, D.D. Dowden. Illustrations by Beverly Magley. Billings, MT: Falcon Press, 1989.

This book is a wonderful field guide for children.

All Ages

THE CARROT SEED, Ruth Krauss. Pictures by Crockett Johnson. New York, NY: Scholastic, 1945.

A little boy has faith that the seed he plants will grow and is rewarded with a giant carrot.

Grades K-2

CHRYSANTHEMUM, Kevin Henkes. New York, NY: The Trumpet Club, 1991.

A little mouse learns that being named for the flower Chrysanthemum isn't so bad after all.

Grades K-1

CORN IS MAIZE, THE GIFT OF THE INDIANS, Aliki Brandenberg. New York, NY: Harper and Row, 1976.

This book describes how corn was found by Indian farmers and how it is grown and used today.

Grades K-2

EACH ORANGE HAD 8 SLICES, Paul Giganti, Jr. Pictures by Donald Crews. New York, NY: The Trumpet Club, 1992.

A counting book of fruit.

Grades K-3

EATING THE ALPHABET, FRUITS AND VEGETABLES FROM A TO Z, Lois Ehlert. New York, NY: The Trumpet Club, 1989.

Eat your way through the alphabet in this book. Includes glossary.

Grades K-2

FACES IN ALL KINDS OF PLACES, Michael Elsohn Ross. Yosemite, CA: Yosemite National Park Association, 1986.

A worm's eye view of flowers. Young readers are introduced to the world of flowers through the eyes of a curious worm.

Grades K-1

THE FLOWER ALPHABET BOOK, Jerry Pallotta. Illustrated by Leslie Evans. Watertown, MA: Charlesbridge Publishing, 1989.

The author, who has written similar books about birds, bugs, the ocean, and trash, uses the alphabet to explore each topic.

Grades K-2

FLOWERS, Rosamund K. Cox. London, UK: Usborne, 1980. (First Nature Books)

Detailed informative illustrations which appeal not only to younger children, but also to older children.

Grades K-4

FLOWERS, Gallimard Jeunesse, Claude Delafosse, and René Mettler. Illustrated by René Mettler. New York, NY: Scholastic, 1991.

In this book young children can peek beneath the petals of a crocus to see what's inside, watch insects pollinate flowers, and follow the movement of sunflowers as they follow the sun. This is a beautiful book.

Grades K-1

GARDEN PARTNERS, Diane Palmisciano. New York, NY: Macmillan Publishing, 1989.

Spring is here, and it is time for grandmother and granddaughter to start their garden. But best of all, it is time for them to share together.

Grades K-2

THE GRANDPA TREE, Michael Donahue. Illustrated by Susan Dorsey. Boulder, CO: Rinehart, Roberts Publications, 1990.

An elementary tale of the life cycle of a tree, from its beginning as a sapling to its demise on the forest floor where it becomes "a home for rabbits, and food for flowers."

Grades K-1

GRANDPA'S TOO-GOOD GARDEN, James Stevenson. New York, NY: The Trumpet Club, 1989.

Overnight, the vegetables grow taller than the house. A zany, outrageously funny tall tale.

Grades K-2

THE GREAT KAPOK TREE: A TALE OF THE AMAZON RAIN FOREST, Lynne Cherry. New York, NY: The Trumpet Club, 1990.

A beautiful book which celebrates the life of the Great Kapok Tree.

Grades K-2

GREEN THUMBS UP, Barbara Taylor. New York, NY: Random House, 1987.

This book explores the science of growing plants. Simple science experiments are included.

Grades K-4

GROWING THINGS, Angela Wilkes. Illustrated by John Shackell. London, UK: Usborne Publishing, 1984.

This book teaches the basic skills a gardener needs to know.

Grades K-4

GROWING VEGETABLE SOUP, Lois Ehlert. New York, NY: Scholastic, 1987.

Through bright illustrations this book traces the beginning of vegetable soup from seed to harvest.

Grades K-1

HOW DO THINGS GROW? Althea. Illustrated by Julie Douglas. Mahwah, NJ: Troll Communications, 1991. (First Science)

This book encourages the reader to explore and understand the science of the way things grow.

Grades 2-3

HOW LEAVES CHANGE, Sylvia A. Johnson. Photographs by Yuko Sato. Minneapolis, MN: Lerner Publications, 1986.

This is an outstanding science book for children and explores the topic, "How do leaves change their color?"

Grades 4-8

JOEY'S CABBAGE PATCH, Harvey Rosenberg. Illustrated by Mary Guariglia. Sebastopol, CA: Go Jolly Publications, 1991.

The reader becomes part of the story by creating their own illustrations. Reproducible.

Grades K-2

JOHNNY APPLESEED, Steven Kellogg. New York, NY: Scholastic, 1988.

This is the story of Johnny Appleseed, a great American frontier hero.

Grades K-2

JUNE 29, 1999, David Wiesner. New York, NY: Clarion Books, 1992.

The giant vegetables in this book are food for the imagination. They can be used as a writing prompt at any grade level.

Grades K-2

A KID'S FIRST BOOK OF GARDENING, Derek Fell. Philadelphia, PA: Running Press, 1989.

A child's book of gardening indoors and out.

Grades 1-3

KIDSGARDENING, A KIDS' GUIDE TO MESSING AROUND IN THE DIRT, Kevin and Kim G. Raftery. Illustrated by Jim M'Guinness. Palo Alto, CA: Klutz Press, 1989.

Growing activities for any place, any season, any climate, any kid! Great gift.

Grades K-2

KILLER PLANTS, Mycol Doyle. Illustrated by Allison Atwill. Los Angeles, CA: Lowell House Juvenile, 1993.

My middle school students love this book.

Grades 4-8

LINNEA IN MONET'S GARDEN, Christina Björk. Pictures by Lena Anderson. New York, NY: R & S Books, 1985.

Linnea visits Monet's garden in Giverny near Paris and traces the artist's life and artwork.

Grades 1-2

MOONSEED AND MISTLETOE, A BOOK OF POISONOUS WILD PLANTS, Carol Lerner. New York, NY: Morrow Junior Books, 1990.

This book is an illustrated guide to the most widely known flowering poisonous plants of temperate North America.

Grades 4-8

MY FIRST GARDEN BOOK, Angela Wilkes. New York, NY: Alfred A.

Knopf, 1992.
This book is full of exciting gardening projects for young children. Clear step-by-step photographs throughout.
Grades K-2

ONE POTATO, Sue Porter. New York, NY: The Trumpet Club, 1989.
This is the classic one potato two potato with a different twist.
Grades K-2

A PICTURE BOOK OF FLOWERS, Theresa Grace. Illustrated by Roseanna Pistolesi. Mahwah, NJ: Troll Communications, 1992.
This book describes and illustrates 20 different flowers.
Grades K-2

PLANT, David Burnie. New York, NY: Alfred A. Knopf, 1989. (Eyewitness Books)
"Like a mini-museum between the covers of a book." A wealth of information about plants.
All Ages

THE PLANT SITTER, Gene Zion. Pictures by Margaret Bloy Graham. New York, NY: Scholastic, 1960.
A little boy becomes a sitter for plants and learns how to propagate them along the way.
Grades K-1

PLANTING A RAINBOW, Lois Ehlert. New York, NY: The Trumpet Club, 1988.
Plant a rainbow of colors. This book illustrates the colors of the rainbow through many different flower varieties.
Grades K-1

PLANTS THAT NEVER EVER BLOOM, Ruth Heller. New York, NY: Scholastic, 1984.
Plants that never ever bloom are called gymnosperms. Beautiful illustrations.
Grades K-2

THE POPCORN BOOK, Tomie de Paola. New York, NY: Scholastic, 1978.
This book has a lot of facts about popcorn.
Grades K-2

PUMPKIN PUMPKIN, Jeanne Titherington. New York, NY: Mulberry Books, 1986.
This is a beautiful book about pumpkins.
Grades K-2

RABBIT SEEDS, Bijou Le Tord. New York, NY: Macmillan Publishing, 1984.
A bunny gardener plans digs, sows, labels, waters, weeds, and harvests her garden.
Grades K-2

THE REASON FOR A FLOWER, Ruth Heller. New York, NY: Grosset & Dunlap, 1983.
Flowers, seeds, and pollination are gorgeously illustrated.
Grades K-4

RED LEAF, YELLOW LEAF, Lois Ehlert. New York, NY: Scholastic, 1990.
This book illustrates the life cycle of a maple tree and the seasons. This is a wonderful book.
Grades K-2

THE ROSE IN MY GARDEN, Arnold and Anita Lobel. New York, NY: William Morrow, 1984.
A cumulative nursery

rhyme set in a garden. Lovely illustrations.

Grades K-3

A SEED IS A PROMISE, Claire Merrill. Illustrated by Susan Swan. New York, NY: Scholastic, 1990.

In every seed there is a promise that a new plant will grow. This book will tell you all about seeds.

Grades 1-2

THIS YEAR'S GARDEN, Mary Szilagyi. Scarsdale, NY: Bradbury Press, 1984.

A rural family plans, plants, and cares for a vegetable garden throughout the seasons.

Grades K-3

THE TINY SEED, Eric Carle. New York, NY: Scholastic, 1987.

The adventures of a seed which is carried by the wind, and planted far away.

Grades K-2

TREE FLOWERS, Millicent E. Selsam. New York, NY: Morrow Publications, 1984.

Twelve flowering trees are described by type and pollination method. Text and labeled drawings follow the growth cycles of apples, magnolias, horse chestnuts, and more.

Grades 3-8

VEGETABLES IN THE GARDEN, Pascale de Bourgoing and Gallimard Jeunesse. Illustrated by Gilbert Houbre. New York, NY: Scholastic, 1989.

All Ages

THE VICTORY GARDEN KIDS BOOK, Marjorie Waters. New York, NY: Houghton Mifflin, 1987.

A beginners guide to growing vegetables.

Grades 2-3

NATURE

ART AND NATURE, AN ILLUSTRATED ANTHOLOGY OF NATURE POETRY, Kate Farrell. Minnetonka, MN: Bullfinch Press, 1992. (Metropolitan Museum of Modern Art)

With 186 poems from around the world, this beautiful anthology takes nature as its inspiration. Organized around the seasons of the year.

All Ages

BOTH SIDES NOW, Joni Mitchell. Illustrated by Alan Baker. New York, NY: Scholastic, 1992.

Joni's song is fancifully illustrated in this delightful book about metamorphosis. Beautiful.

Grades K-1

CHILDREN OF THE DUST BOWL, THE TRUE STORY OF THE SCHOOL AT WEEDPATCH CAMP, Jerry Stanley. New York, NY: Crown Publishers, 1992.

This book begins on the drought-stricken plains of Oklahoma and describes the journey west of the "Oakies" of the 1930s. A memorable glimpse into a neglected period of American history, this is a wonderful, historical recording of a child's view of the "Dirty Thirties."

All Ages

EARTH SONGS, Myra Cohn Livingston and Leonard Everett Fisher. New York, NY: Scholastic, 1986.

From the desert to the ocean, from the high mountains to the tundra, these songs poetically evoke the many different environments on earth.

Grades K-4

FULL WORM MOON, Margo Lemieux. Illustrated by Robert Andrew Parker. New York, NY: Tambourine Books, 1994.

This is an original Algonquian family story. The Full Worm Moon occurs in March when the sunlight is getting stronger. You can tell the worms are waking up when you find their castings on the surface of the earth.

Grades K-2

GROWING COLORS, Bruce McMillan. New York, NY: Lothrop, Lee and Shepard Books, 1988.

Is a bean always green, or might it be purple? The author takes his camera into the garden and orchard to discover color as it grows in nature. A wonderful book for emergent readers.

Grades K-2

IN MY MOTHER'S HOUSE, Ann Nolan Clark. Illustrated by Velino Herra. New York, NY: Viking Press, 1941.

A old story with a timeless message which celebrates the harvest and the family.

Grades K-2

THE OCEAN ALPHABET BOOK, Jerry Pallotta. Illustrated by Frank Mazzola, Jr. Watertown, MA: Charlesbridge Publishing, 1989.

The author, who has written similar books about birds, bugs, flowers, and trash, uses the alphabet to explore each topic.

Grades K-2

THE REASONS FOR SEASONS, Linda Allison. Covelo, CA: Yolla Bolly Press, 1975. (A Brown Paper School Book)

Science activities and interesting facts for winter, spring, summer, and fall.

Grades 4-8

SCARECROW! Valerie Littlewood. New York, NY: Dutton Children's Books, 1992.

This book gives us a glimpse of the myths and magic surrounding scarecrows.

Grades 3-5

THE SCARECROW BOOK, James Giblin and Dale Ferguson. New York, NY: Crown Publishers, 1980.

The history and multicultural variations of scarecrows, illustrated with interesting photographs.

Grades 2-6

SPRING IS HERE, Taro Gomi. New York, NY: The Trumpet Club, 1989.

This book traces the seasons through the growth of a calf.

Grades K-1

A WALK IN THE RAIN FOREST, Kristin J. Pratt. Nevada City, CA: Dawn Publications, 1992.

Written by a 15-year old environmentalist and high-school student. Beautiful informative book written with sensitivity to the

plight of the rain forests.
All Ages

WELCOME TO THE GREENHOUSE, Jane Yolen. Illustrated by Laura Regan. New York, NY: G. P. Putnam, 1990.
A beautiful book about the rain forest:
"Welcome to the green house
Welcome to the hot house
Welcome to the land of warm, wet days."
Grades K-2

WHERE THE FOREST MEETS THE SEA, Jeannie Baker. New York, NY: Scholastic, 1987. (Environment)
A boy and his father spend a day in the Daintree rain forest of Australia.
Grades K-2

WILD, WILD SUNFLOWER CHILD ANNA, Nancy White Calstrom. Illustrated by Jerry Pinkney. New York, NY: Scholastic, 1987.
Anna explores the wild environment of her garden.
Grades K-2

VALUES

THE CORN GROWS RIPE, Dorothy Rhoads. Illustrated by Jean Charlot. New York, NY: Puffin Books, 1993. (A Newbery Honor Book)
When a boy's father is injured in an accident, who will harvest the corn and please the Mayan gods? This is the wonderful story rich with Mayan folkways.
All Ages

THE EMPTY POT, Hitz Demi. New York, NY: The Trumpet Club, 1990.
This story takes place in China. A little boy learns the importance of honesty.
Grades K-2

THE FLOWER FAIRY, Frank Asch and Vladimir Vagin. New York, NY: Scholastic, 1993.
An exquisitely illustrated fable of love between father and son and the problems that ensue when nature and human nature fall out of balance.
Grades K-2

JUST A DREAM, Chris Van Allsburg. New York, NY: Houghton Mifflin, 1990.
A very careless boy learns the importance of recycling. Great book!
Grades K-2

THE LEGEND OF THE BLUEBONNET, AN OLD TALE RETOLD, Tomie de Paola. New York, NY: Scholastic, 1990.
In this Comanche legend a young girl sacrifices her most prized possession to save her people.
Grades K-2

A MIGRANT FAMILY, Larry Dane Brimner. Minneapolis, MN: Lerner Publications, 1993.
Twelve-year-old J. Medina and his family live in a makeshift shack in San Diego. In this honest, provocative photographic essay Larry Dane Brimner shows how one migrant family gets along.
Grades 4-8

MISS RUMPHIUS, Barbara Cooney. New York, NY: The Trumpet Club, 1982.

Alice Rumphius wanted to travel the world when she grew up, and then to live by the sea, just as her grandfather had done.

Grades K-2

THE SEASONS SEWN, A YEAR IN PATCHWORK, Ann Whitford Paul. Illustrated by Michael McCurdy. Lake Oswego, OR: Browndeer Press, 1996.

This beautiful book shows how the names of wonderful quilt patterns came to be and teaches us the way of life in the 19th century.

All Ages

THE SECRET GARDEN, Frances H. Burnett. New York, NY: Playmore, 1994.

The beautiful tale of a young girl's experience after her parents die.

All Ages

TSUBU THE LITTLE SNAIL, Carol Ann Williams. Illustrated by Tatsuro Kiuchi. New York, NY: Simon and Schuster, 1995. (Books for Young Readers)

In this authentic Japanese folktale, a rice farmer and his wife pray for a child. They are sent a snail by the Water God instead.

Grades K-4

WORKING COTTON, Sherley Anne Williams. Illustrated by Carole Byard. New York, NY: Harcourt Brace Jovanovich, 1992.

A child's experience in the cotton fields of Fresno.

Grades K-2

YONDER, Tony Johnston. Pictures by Lloyd Bloom. New York, NY: Dial Books, 1990.

"Yonder is the farmer on the hills that roll forever . . . ." Johnston's words capture the cyclical pattern of a farming way of life.

Grades K-1

THE YUCKY ALPHABET BOOK, Jerry Pallotta. Illustrated by Ralph Masiello. Watertown, MA: Charlesbridge Publishing, 1990.

The author, who has written similar books about birds, bugs, flowers, and the ocean, uses the alphabet to explore each topic.

Grades K-2

II. Curriculum and Teacher Resources

ANIMALS

ANIMAL LIFE CYCLES, Jo E. Moore and Joy Evans. Poster art by Leslie Tryon. Monterey, CA : Evan-Moor, 1986. (Science Mini-Units)

These 16-sheet units on birds, amphibians, and insects include posters and reproducible pages with word search games and crossword puzzles.

Grades 1-3

BIRDWISE: FORTY FUN FEATS FOR FINDING OUT ABOUT OUR FEATHERED FRIENDS, Pamela M. Hickman. Illustrated by Judie Shore. New York, NY: Addison-Wesley, 1989.

This is a great kids introduction to our feathered

friends.

Grades 2-7

<u>THE BUG BOOK: HARMLESS INSECT CONTROLS</u>, Helen and John Philbrick. Illustrated by Cathy Baker. Pownal, VT: Storey Communications, 1974.

This book tells you how to identify insects that are harmful or helpful in your garden. There is a recipe section for creating concoctions that will control insect pests. A helpful guide for gardeners who do not want to use pesticides.

<u>BUGPLAY: ACTIVITIES WITH INSECTS FOR YOUNG CHILDREN</u>, Leon H. Burton. New York, NY: Addison-Wesley, 1989.

Guide your students through a hands-on exploration of the insect world. Poems and blackline masters reinforce the unique features of 17 insect orders.

<u>BUGWISE</u>, Pamela M. Hickman. Illustrated by Judie Shore. Reading, MA: Addison-Wesley, 1991.

Thirty incredible insect investigations and arachnid activities.

Grades 2-7

<u>BUTTERFLIES ABOUND, A WHOLE LANGUAGE APPROACH TO BUTTERFLY SCIENCE</u>, Seddon K. Beatty. Reading, MA: Addison-Wesley, 1993.

This book is a treasure of resources.

<u>BUTTERFLY</u>, Robert Michael Pyle. Illustrated by Sarah Anne Hughes. New York, NY: Houghton Mifflin, 1983.

This is a wonderful text and one of only a few that promote watching rather than the collecting these beautiful insects. I cannot recommend this book highly enough. Be sure to include it in your resource library. It covers everything from ecology to photography.

<u>THE BUTTERFLY BOOK, AN EASY GUIDE TO BUTTERFLY GARDENING, IDENTIFICATION, AND BEHAVIOR</u>, Donald Stokes, Lillian Stokes, and Ernest Williams. New York, NY: Little, Brown, 1991.

This book has stunning photographs of butterflies in various life stages. It tells you how to design butterfly habitats and much more. These authors also produce nature guides, wildflower books, and bird books.

<u>THE CALIFORNIA FRESHWATER SHRIMP PROJECT</u>, Laurette Rogers. Berkeley, CA: Heyday Books, 1996.

In this example of environmental project-based learning with funding from the Center for Ecoliteracy in Berkeley, The author describes a style of project-based learning where students assume the responsibility for their own learning, take risks, and do real work. Project-based learning can be applied to many situations. I highly recommend this book to teachers, parents and administrators.

<u>COLOR HANDBOOK OF GARDEN INSECTS</u>, Anna Carr. Emmaus, PA: Rodale Press, 1979.

Over 300 full color photographs of insects.

COMPOST CRITTERS, Bianca Lavies. New York, NY: Dutton Children's Books, 1993.

This is a great book. The photographs of organisms found in compost piles are the best I've seen. A "must have."

CREEPY CRAWLIES: A THEMATIC UNIT, Mary Ellen Sterling. Illustrated by Paula Spence. Huntington Beach, CA: Teacher Created Materials, 1990.

Reproducible; a whole language approach to a thematic unit about insects.

Grades K-2

CREEPY CRAWLIES AND THE SCIENTIFIC METHOD, Sally Stenhouse Kneidel. Golden, CO: Fulcrum Publishing, 1993.

This book contains over 100 hands on science experiments for children.

CREEPY CRAWLIES FOR CURIOUS KIDS, Lynn Ransford. Huntington Beach, CA: Teacher Created Materials, 1986.

These are teacher-generated materials for the classroom which have blackline masters you can reproduce. This volume covers earwigs, bees, spiders, earthworms, and more.

CRITTERS. Fresno, CA: AIMS Education Foundation, 1990. (Life Science)

Topics include insects, spiders, snails, meal worms, and more. Beautiful illustrations of earthworms.

EARTHWORMS, Robert C. Knott, Kimi Hosoume, and Lincoln Bergman. Berkeley, CA: Lawrence Hall of Science, University of California, 1989. (Great Explorations in Math and Science, GEMS)

This is a curriculum guide for teaching earthworm science. It provides an outline and some reproducible masters for you to guide your students through a study of earthworms.

Grades 6-10

EARTHWORMS, DIRT, AND ROTTEN LEAVES, Molly McLaughlin. Illustrated by Robert Shetterly. New York, NY: Avon Books, 1986, 1990.

This was one of the first titles on earthworms I read. Both editions are now out of print. However, they are well organized, give us a lot of information about our segmented friends, and are worth looking for in libraries and wherever old books are for sale.

THE ENCYCLOPEDIA OF BUTTERFLIES, John Feltwell. New York, NY: Prentice Hall, 1993. (General Reference)

This is a great reference book with beautiful colored plates that illustrate more than 1,000 butterflies from around the world.

GARDEN BIRDS OF AMERICA, George H. Harrison. Minocqua, WI: Willow Creek Press, 1997. Video.

This is a gallery of the song birds you can invite to your garden. Travel across America and learn about birds and the gardens they like to visit and why. I use this video in class. *Call* 1-800-850-9453

THE HANDBOOK FOR BUTTERFLY WATCHERS, Robert Michael Pyle. Illustrated by Sarah Anne Hughes. Boston, MA: Houghton Mifflin, 1992.

Great for identification.

INSECTS AND OTHER CRAWLERS, Joan Westley. Mountain View, CA: Creative Publications, 1988. (Windows on Beginning Science)

Contains simple lessons for grades K-2 that can be taught by parents and cross-age tutors.

MONARCHS, Kathryn Lasky. Photography by Christopher G. Knight. New York, NY: Harcourt Brace, 1993.

Learn about Monarchs.

WORMANIA, Mary Appelhof, Billy B. and the Brennan Kids. Kalamazoo, MI: Flowerfield Enterprises, 1995. Video, 26 minutes.

This is a "must have" for teaching worm science. The footage of baby worms is outstanding and is highly recommended! Although geared for younger children, I show this film to my middle school students because of the content and photography. Don't buy this video if the euphemism "worm poop" offends you.

Grades 4-8

PLANTS

THE AMAZING APPLE BOOK, Paulette Bourgeois. Illustrated by Linda Hendry. New York, NY: Addison-Wesley, 1990.

Show off with dozens of fun projects using apple science.

Grades 1-5

THE AMAZING POTATO BOOK, Paulette Bourgeois. Illustrated by Linda Hendry. New York, NY: Addison-Wesley, 1991.

Shows you lots of ways to have fun with potatoes. Map how Spanish explorers brought potatoes from South America to Europe.

Grades 2-6

BETTER HOMES AND GARDENS: COMPLETE GUIDE TO GARDENING. Seattle, WA: Multi-Com Publishing Co, 1994. CD-ROM.

This is a good reference guide on gardening techniques; it also has an almanac where you can keep a gardener's notebook.

BIO-DYNAMIC AGRICULTURE: AN INTRODUCTION, Herbert H. Koepf and Wolfgang Schaumann. Hudson, NY: Anthroposophic Press, 1976.

An introduction to biodynamic gardening.

BLUE CORN & SQUARE TOMATOES: UNUSUAL FACTS ABOUT COMMON GARDEN VEGETABLES, Rebecca Rupp. Pownal, VT: Storey Communications, 1987.

This book has unusual facts about common vegetables. Why are peppers hot? Why are carrots always orange? Find out when you read this text written by Rebecca Rupp who holds a Ph.D. in biology. It also includes recipes. Useful when integrating history with plant use.

THE BOOK OF GARLIC, Lloyd J. Harris. Reading, MA: Addison-Wesley, 1979.

This publication tells you everything you always wanted to know about garlic and more. It has remedies, recipes, folklore, history, and new contributions by the Lovers of the Stinking Rose. A very entertaining book.

BOTANY FOR ALL AGES: DISCOVERING NATURE THROUGH ACTIVITIES FOR CHILDREN AND ADULTS, Jorie Hunken. Old Saybrook, CT: The Globe Pequot Press, 1993.

This publication can be used by parents as well as teachers or anyone wishing to learn about the environment with children. It is a guide that contains lesson plans covering a range of botanical subjects. From the study of the parts of plants to the importance of preserving genetic diversity, this is a good tool for outdoor explorations with children.

THE BUDDING BOTANIST: INVESTIGATIONS WITH PLANTS, Evalyn Hoover, et al. Fresno, CA: AIMS Education Foundation, 1993.

This book is concerned with the investigation of seed plants and a beginning knowledge of the source of plants, their structures and their economic importance.

Grades 3-6

BUFFALO BIRD WOMAN'S GARDEN, Gilbert L. Wilson. St. Paul, MN: Minnesota Historical Society Press, 1987.

"This that I now tell is as my mothers do, or did myself, when I was young. My mothers were industrious women, and our family had always good crops; and I will tell you now how the women of my father's family cared for their fields, as I saw then, and helped them,"writes Buffalo Bird Woman, an Hidatsa Indian and excellent gardener. For me this is a very special book that provides us with methods we can use today. Useful when integrating history with plant use.

THE CAPTURED HARVEST, CREATING EXQUISITE OBJECTS FROM NATURE, Terence Moore. North Pomfret, VT: Trafalgar Square Publishing, 1993.

This book is a step-by-step guide to creating objects made from things found in nature. The author uses seeds, cones, leaves, twigs, and more to create beautiful displays of nature. It includes over 50 projects.

CARROTS LOVE TOMATOES, Louise Riotte. Pownal, VT: Storey Communications, 1975.

This book teaches you about companion planting. Hence, the title of the book. It discusses the relationship between plants, and guides successful planting of vegetables in your garden.

THE COLOR DICTIONARY OF FLOWERS AND PLANTS FOR HOME AND GARDEN, Roy Hay and Patrick M. Synge. New York, NY: Crown Publishing, 1982.

This is *great* reference book with 2048 illustrations in full color.

COMMON WEEDS COLORING BOOK, Stefen Bernath. New York, NY: Dover Publications, 1976.

The 45 weeds illustrated

here are typical important weeds, likely to be found almost anywhere in the United States. Full captions identify them, with both botanical and popular names.

All Ages

<u>COMPLETE GARDEN GUIDE TO THE NATIVE SHRUBS OF CALIFORNIA</u>, Glenn Keator. San Francisco, CA: Chronicle Books, 1994.

This is the only publication I know of devoted to this subject. It discusses soil science, climate, topography, transplanting, and pruning. It also is full of interesting facts and tips for those who are trying to build landscapes reflecting the bioregion in which they live. Great resource.

<u>CREATING A GARDEN FOR THE SENSES</u>, Jeff Cox. Photographs by Jerry Pavia. New York, NY: Abbeville Publishers, 1993.

This is a beautiful resource book that is divided into gardens for sight, smell, touch, sound, taste and the sixth sense which is, "the key to living an artful life . . . intuition." I like the way this book is organized.

<u>CULTIVATING A CHILD'S IMAGINATION THROUGH GARDENING</u>, Nancy Allen Jurenka and Rosanne J Blass. Englewood, CO: Teacher Ideas Press, 1996.

Here are 45 lessons that promote literacy and fun!

<u>DICTIONARY OF HORTICULTURE</u>, National Gardening Association. New York, NY: Penguin Books, 1994.

This is a good reference book.

<u>DIGGING DEEPER, INTEGRATING YOUTH GARDENS INTO SCHOOLS AND COMMUNITIES</u>, Joseph Kiefer and Martin Kemple. Foreward by Alice Waters. Montpelier, VT: Common Roots Press, 1998.

Published as a partnership by Food Works, the American Community Gardening Association, and Let's Get Growing, this is a great resource and comprehensive guide if you are beginning a garden project from the ground up. It has practical ideas for educators and volunteers alike.

<u>EDIBLE WILD PLANTS, A NORTH AMERICAN FIELD GUIDE</u>, Thomas S. Elias and Peter A. Dykeman. New York, NY: Sterling Publishing, 1990. (An Outdoor Life Book)

This book is a guide for harvesting and preparing over 200 plants collected from the wild. It has maps and nearly 400 photographs, some in color, to help identify each plant.

<u>THE ENCYCLOPEDIA OF ORGANIC GARDENING</u>, Organic Gardening® Magazine staff. Revised edition. Emmaus, PA: Rodale Press, 1978.

A wealth of alphabetized information about specific crops and horticultural practices for the organic gardener, from composting to organic pest control.

<u>EVERLASTING FLORAL GIFTS</u>, Rob Pulleyn and Claudette Mautor.

New York, NY: Sterling Publishing, 1990.

During the school year, I often teach a gardening elective to 6th, 7th, and 8th graders. When the weather turns cold and wet, and we must be inside, I gather all of the dried flowers and grasses we have grown or collected for the students to use in nature craft projects. This book provides more than 100 creative ideas.

<u>FAVORITE WILDFLOWERS COLORING BOOK</u>, Ilil Arbel. New York, NY: Dover Publications, 1991.

This book is great to use for its blackline masters when teaching wildflower identification.

<u>FLORAL PORTRAITS OF CALIFORNIA FROM RANCHO SANTA ANA BOTANIC GARDEN</u>, B.E. Busenberg. Drawings by Lois Martin. Claremont, CA: Rancho Santa Ana Botanic Garden, 1991.

This booklet has beautiful line drawings of many of California's native flowers.

<u>FLOWER DRYING WITH A MICROWAVE: TECHNIQUES AND PROJECTS</u>, Titia Joosten. Asheville, NC: Altamont Press, 1988.

I have used this book a lot when I am doing nature crafts with students. It is great when you need to dry a lot of flowers in a short amount of time. This method works well with roses.

<u>GARDEN FLOWERS COLORING BOOK</u>, Stefen Bernath. New York, NY: Dover Publications, 1975.

This text gives you good blackline masters of common garden flowers.

<u>THE GARDENER'S BOOK OF COLOR</u>, Andrew Lawson. Pleasantville, NY: Reader's Digest Association, 1996.

This book has beautiful inspiring photographs and explores the use of color in the garden. Includes diagrams of the plantings photographed in the book.

<u>GARDENER'S LATIN</u>, Bill Neal. Chapel Hill, NC: Algonquin Books of Chapel Hill, 1992.

This book was a "must have" for me when teaching the Latin names of plants. Very useful.

<u>GARDENING WITH CHILDREN, A BOOK OF PATTERNS</u>, Fran Edge. Illustrations by Melissa Johnson. Eureka, CA, n.d.

This pattern book was developed for anyone wishing to garden with small children. Nice illustrations! *Write to* Fran Edge, 1604 K St., Eureka, CA 95501.

Grades Pre-1

<u>GARDENS OF THE WORLD</u>, hosted by Audrey Hepburn. N.p., Perennial Productions, 1993. Video.

This is a series of six different types of gardens found throughout the world. Included are Rose Gardens, Tulip Gardens, Flower Gardens, Formal Gardens, Country Gardens, and Public Gardens. These videos also incorporate geography and history. I use these in class.

<u>GET READY, GET SET, GROW: A KID'S GUIDE TO GOOD GARDENING.</u>

Brooklyn, NY: Brooklyn Botanical Garden, 1986? Video, 15 minutes.

A 15-minute video cassette which shows children gardening.

GROW LAB, A COMPLETE GUIDE TO GARDENING IN THE CLASSROOM, Eve Pranis and Jack Hale. Burlington, VT: National Gardening Association, 1988.

This book is designed to be used with the National Gardening Association's Grow Lab Indoor Garden.

THE GROWING CLASSROOM, Roberta Jaffe and Gary Appel. New York, NY: Addison-Wesley, 1982.

A detailed program that combines gardening and nutrition in 13 complete hands-on units. This curriculum can also be used with 7th and 8th graders. Life Lab is currently working on a middle-school curriculum guide. YEA!

Grades 4-8

GROWING THINGS, Angela Wilkes. Illustrated by John Shakell. London, UK: Usborne Publishing, 1984.

This book has some great activities for kids.

HERBS, Holly Smimizu. Silver Springs, MD: Audio Visual Artists Productions, 1995. Video.

This video provides a tour of several beautiful gardens. The narrator is an expert when it comes to herbs. I use this in class.

HIDDEN STORIES IN PLANTS, Anne Pellowski. New York, NY: Macmillan Publishing, 1990.

This reference book contains myths, legends, tales, and folktales. It also describes how to use plants to make ornaments, toys, disguises, dolls, and musical instruments.

HOW FLOWERS WORK, A GUIDE TO PLANT BIOLOGY, Bob Gibbons. Poole, Dorset, UK: Blandford Press, 1984.

This guide shows how the flowering plant fits into the plant kingdom as a whole, how each separate part of the plant functions, and explains the reasons for the production of flowers in their numerous different forms.

Grades 7-12

HOW TO GROW BULBS, Sunset Books and Magazine. Menlo Park, CA: Lane Publishing, 1973.

Very useful when growing bulbs.

HOW TO GROW MORE VEGETABLES, John Jeavons. Berkeley, CA: Ten Speed Press, 1982.

This man is a local and works with many Third World countries. Very interesting and informative.

HOW TO KNOW THE WESTERN TREES, Harry J. Baerg. Dubuque, IA: William Brown, 1955.

Very good for identification.

THE ILLUSTRATED BOOK OF TREES & SHRUBS, Eleanor Lawrence. New York, NY: Gallery Books, 1985.

This has beautiful drawings.

AN ILLUSTRATED GUIDE TO ATTRACTING BIRDS, Sunset Books and Magazine. Menlo Park, CA, 1990.

Learn how to attract birds to your yard.

IN FULL VIEW, THREE WAYS OF SEEING CALIFORNIA PLANTS, Glenn Keator, Linda Yamane, and Ann Lewis. Berkeley, CA: Heyday Books, 1995.

This book has us look at native plants through three perspectives giving us a full view of the plants of California.

KASHAYA POMO PLANTS, Jennie Goodrich, Claudia Lawson, and Vana Parrish Lawson. Berkeley, CA: Heyday Books, 1980.

This book describes 150 common plants of coastal California, and describes how each plant is used according to tribal custom.

A KID'S FIRST BOOK OF GARDENING, Derek Fell. Philadelphia, PA: Running Press, 1989.

Great for beginners.

KILLER PLANTS, Mycol Doyle. Illustrated by Allison Atwill. Los Angeles, CA: Lowell House Juvenile, 1993.

My middle-school students always like this book. So far I'm still alive.

LEARNING ABOUT PLANTS, Jo Ellen Moore and Joy Evans. Monterey, CA: Evan-Moor Corp., 1987.

Reproducible worksheets; this book covers the structure and characteristics of plants.

Grades 1-2

LET'S GROW! Linda Tilgner. Pownal, VT: Storey Communications, 1988.

This book covers working with the handicapped and very small children through 72 gardening adventures.

MEDICINAL PLANTS COLORING BOOK, Ilil Arbel. Mineola, NY: Dover Publications, 1993.

This informative coloring book features 44 accurate drawings by a botanical illustrator.

NATIONAL GARDENING, A MONTHLY MAGAZINE FOR GARDENERS. Burlington, VT: National Gardening Association.

Basic gardening tips, creative gardening ideas, and addresses for seed exchanges. Great for letter writing by elementary school children! *Write* NGA Member Subscription Service, Depot Sq., Peterborough, NH 03458; $15.00 subscription includes membership to NGA.

THE NATURAL HABITAT GARDEN, Ken Druse and Margaret Roach. New York, NY: Clarkson Potter, 1994.

Beautiful book that teaches you how to put together a habitat garden to entice creatures into your world.

NATURE PRINTING WITH HERBS, FRUITS, & FLOWERS, Laura Donnelly Bethmann. Pownal, VT: Storey Communications, 1996.

Beautiful book that teaches you how to use flowers fruits and herbs to create art.

ORIENTAL GARDENING, Kate Jerome. New York, NY: Pantheon Books, 1996.

Beautiful book which looks at Oriental gardening.

PAPER FLOWERS, Ray Gibson. Tulsa, OK: Usborne Publishing, 1994.

Good craft ideas.

PEOPLE, PLANTS, AND PLACES, Julian Agyeman. Devon, England: Southgate Publishers, 1995. (Learning Through Landscapes)

Cultural ecology uses plants to provide a window when looking at diversity. Such gardens are valuable to every school.

PLANT, David Burnie. New York, NY: Alfred A. Knopf, 1989. (Eyewitness Books)

I love this series of books for their bold color and information.

PLANT AND FLORAL WOODCUTS FOR DESIGNERS & CRAFTSMEN, Theodore Menten. New York, NY: Dover Publications, 1974.

This is a great book to have when teaching science or art. Beautiful reproducible prints.

PLANTS USED BY THE INDIANS OF MENDOCINO COUNTY, CALIFORNIA, V.K. Chestnut. Ukiah, CA: Mendocino County Historical Society, 1974. (Contributions from the U.S. National Herbarium, Vol. VII)

The title is somewhat misleading as California Indians, foreign to Mendocino County, were brought to the Round Valley Indian Reservation after its establishment in 1856. Most of the plants covered in this book are found in other areas of the western United States.

THE PLEASURES OF HERBS, Phyllis Shaudys. Pownal, VT: Storey Communications, 1986.

I have used this book for years. It is organized around the months of the year and has activities for each month.

PRIMARILY PLANTS. Fresno, CA: AIMS Educational Foundation, 1990.

Contains 22 investigations that cover all major K-3 concepts identified in the "Biological Science: Plants" section of the California Science Framework.

Grades K-3

SEEDS AND WEEDS, Joan Westley. Mountain View, CA: Creative Publications, 1988. (Windows on Beginning Science)

Contains simple lessons that can be taught by parents, and cross-age tutors.

Grades K-2

SEEDS OF CHANGE, FIVE PLANTS THAT TRANSFORMED MANKIND, Henry Hobhouse. New York, NY: Harper & Row, 1985. (Perennial Library)

Five plants—quinine, sugar cane, tea, cotton, and the potato—have, in effect, been powerful historical catalysts. This book traces their origin and impact.

SLEEPING WITH A SUNFLOWER, A TREASURY OF OLD-TIME GARDENING LORE, Louise Riotte. Pownal, VT: Storey Communications, 1987.

Folkore; a treasury of old time gardening lore.

TAYLOR'S GUIDE TO BULBS. Boston, MA: Houghton Mifflin, 1986.

Based on *Taylor's Encyclo-*

pedia of Gardening, this is a good guide for bulb growing.

THEME GARDENS, Barbara Damrosch. Illustrated by Karl W. Stuecklen. New York, NY: Workman Publishing, 1982.
This book presents different themes you can create in a garden from Shakespeare to old roses. Great book.

THE VISUAL DICTIONARY OF PLANTS. New York, NY: Dorling Kindersley, 1992. (Eyewitness Visual Dictionaries)
I love this series.

WILDFLOWER FOLKLORE, Laura C. Martin. Old Saybrook, CT: Globe Pequot Press, 1984.
This is a wonderful book full of folklore.

WILDFLOWERS. Alexandria, VA: Time, 1989. (The Time Life Gardener's Guide)
This volume provides a wealth of information on growing wildflowers of all types, from a wide range of climates.

WILDFLOWERS, Roger Tory Peterson and Frances Tenebaum. New York, NY: Houghton Mifflin, 1982.
This is a field guide coloring book with very good drawings of 146 wildflowers for use in the classroom.

WILDFLOWERS ACROSS AMERICA, Lady Bird Johnson and Carlton B. Lees. New York, NY: Abbeville Press, 1993.
A beautiful American journey with Lady Bird Johnson.

THE WORLD IN YOUR GARDEN. Washington, DC: National Geographic Society, 1957.
I found this book in an old book store in Eureka, CA. If you can find it get yourself a copy for a look at the origins of fruits and veggies.

THE WREATH BOOK, Rob Pulleyn. New York, NY: Sterling Publishing, 1988.
This is a wonderful book and has a lot of ideas on how to use dried flowers in wreath making.

NATURE

THE AMAZING DIRT BOOK, Paulette Bourgeois. New York, NY: Addison-Wesley, 1990.
The Amazing Dirt Book gets "down and dirty" with two dozen earthy experiments.
All Ages

BEYOND THE BEAN SEED: GARDENING ACTIVITIES FOR GRADES K-6, Nancy Allen Jurenka and Rosanne J Blass. Englewood, CO: Teacher Idea Press, 1996.
There are a lot of hands-on activities in this book. Topics range from butterflies to word play.

BUTTERFLY GARDENING. New York, NY: Audubon Society, Master-Vision, 1995. Video.
Know what plants to plant in your garden to attract over 60 different species of butterflies. I use this video in class.

EDUCATION GOES OUTDOORS, Frank D. Johns. Menlo Park, CA: Addison

Wesley, 1986.

This volume is full of learning experiences that teach students basic skills through direct contact with the natural world.

Grades K-9

ENCHANTED CIRCLES, Elizabeth Jane Lloyd. New York, NY: Simon and Schuster, 1990.

This beautiful book traces the history of circles and rings and their power for mankind since the earliest of times. It explores the importance and significance of the circle as a symbol. The author gathered examples of wreaths from many different countries. From French herbal wreaths to English corn circles, this book is a wonderful inspiration for making wreaths.

GARDEN INSECT, DISEASE, AND WEED IDENTIFICATION GUIDE, Miranda Smith and Anna Carr. Emmaus, PA: Rodale Press, 1988.

Good identification guide.

GARDEN PROBLEM SOLVER, Menlo Park, CA: Sunset Publishing, 1996. CD-ROM.

This CD is a guide to creating a healthy garden. It identifies 400 insects, diseases, weeds and problems you might find in the garden.

HANDS-ON NATURE, Jenepher Lingelbach. Woodstock, VT: Vermont Institute of Natural Science, 1986.

Explore adaptations, habitats, cycles and designs of nature with this guide.

Grades Pre-8

HANDS-ON SCIENCE, TEN THEMES FOR A WHOLE YEAR, Kay Ballantyne, et al. Monterey, CA: Evan-Moor Corp., 1993.

Each unit contains a teacher resource page, student lab book, and parent letter.

Grades 1-2

IN MY BACKYARD, Jo Moore and Joy Evans. Monterey, CA: Evan-Moor Corp., n.d.

Color and learn backyard science facts. Reproducibles.

Grade K

LITERATURE FOR SCIENCE AND MATHEMATICS. Sacramento, CA: California Department of Education, 1993. (Science and Environmental Education Unit)

An annotated bibliography of science and mathematics related literature.

Grades K-12

LIVING IN THE ENVIRONMENT, G. Tyler Miller, Jr. 6th ed. Belmont, CA: Wadsworth Publishing, 1990.

This book is a teacher text and is a good resource when teaching students about the environment. Subject matter is divided into categories: Water Resources, Food Resources, Soil Resources, etc. It offers a way to teach science content under the umbrella of a big idea.

NATURECRAFTS, 50 EXTRAORDINARY GIFTS & PROJECTS, STEP BY STEP, Gillian Souter. New York, NY: Crown Trade Paperbacks, 1996.

This is full of fun activities.

NATUREWATCH, EXPLORING NATURE WITH YOUR CHILDREN, Adrienne Katz. New York, NY: Addison-Wesley, 1986.

Offers your students dozens of ways to explore and learn from their natural environment.

All Ages

OF CABBAGES AND CHEMISTRY, Jacqueline Barber. Berkeley, CA: Lawrence Hall of Science, University of California, 1989. (Great Explorations in Math and Science, GEMS)

Contains a series of engaging activities for you and your students to explore acids and bases using the special indicator properties of red cabbage juice.

Grades 4-8

ONE SMALL SQUARE, BACKYARD. Los Angeles, CA: Virgin Sound and Vision, 1995. CD-ROM.

Shows you how to take a closer look at the earth. This CD explores plants, soil and animals.

OUR WONDERFUL WORLD. Fresno, CA: AIMS Education Foundation, 1987.

An environmental studies booklet. Topics include: air, water, transpiration, soil, plants, animals and insects.

Grades 4-9

ROCKS, SAND, AND SOIL, Joan Westley. Sunnyvale, CA: Creative Publications, 1988. (Windows on Science)

Contains simple lessons that can be taught by parents, and cross-age tutors.

Grades K-2

SCARECROW! Valerie Littlewood. New York, NY: Dutton Children's Books, 1982.

It isn't easy to find anything on scarecrows. This text has a lot of interesting facts in it.

SEEDS OF CHANGE, THE STORY OF CULTURAL EXCHANGE AFTER 1492, Sharryl Davis Hawke and James E. Davis. Menlo Park, CA: Addison-Wesley, 1992.

This publication introduces students in grades 5-8 to cultural and biological interchanges. Student's workbook and teacher's guide available.

TINA'S SCIENCE NOTEBOOK, Tina DeCloux and Rosanne Werges. Illustrations by Tara Sullivan. San Rafael, CA: Symbiosis Books, 1985.

Topics include: birds, seeds, predator and prey and more!

Grades K-2

WEATHER WATCH, Valerie Wyatt. Illustrated by Pat Cupples. Menlo Park, CA: Addison-Wesley, n.d.

Showers you with wacky weather wisdom, a flurry of fun facts and two dozen projects for both rainy and sunny days.

Grades 2-7

THE WEB OF LIFE, Fritjof Capra. New York, NY: Anchor Books, 1996.

This book, based on ten years of research, gives us a new perspective on the nature of life. Capra presents a new understanding of the science of living systems. I highly recommend this book.

WETLANDS PLANTS AND ANIMALS COLORING BOOK, Annika

Bernhard. New York, NY: Dover Publications, 1994.

This book provides blackline masters of wetlands.

VALUES

THE AMAZING PAPER BOOK, Paulette Bourgeois. Illustrated by Linda Hendry. New York, NY: Addison-Wesley, 1989.

More than two dozen eye-opening ways to use paper.

Grades 2-6

BACKYARD COMPOSTING: YOUR COMPLETE GUIDE TO RECYCLING YARD CLIPPINGS, John W. Roulac. Ojai, CA: Harmonious Press, 1992.

Creative methods to dispose of green wastes.

THE COMPLETE HOME COMPOSTING VIDEO GUIDE, Howard Stenn. N.p., Venal Productions, 1993. Video.

Howard is an ecological horticulturist and gives an overview of composting: yard waste, compost bins, worm composting, trouble-shooting, pest control, the five essentials, sod composting, hot composting, compost uses, compost tools and more. This is a very informative video. I use this in class.

COMPOSTING, A STUDY OF THE PROCESS AND ITS PRINCIPLES, Clarence G. Golueke. Emmamus, PA: Rodale Press, 1975.

This is a teacher resource for the history of composting and its process and principles.

COMPOSTING SOURCE SEPARATED ORGANICS, BioCycle Magazine staff. Emmaus PA: JG Press, 1994.

This is an in-depth look at composting.

DO THE ROT THING. Alameda County, CA: Waste Management Authority and Source Reduction and Recycling Board, 1997. Video.

My friends in Alameda have been way ahead of the game when it comes to teacher and student education. This is a wonderful film which demonstrates the "how to" of composting. I like the photography and use this video with classes. *See also* this website http://www.stopwaste.org for more information.

EARTH CHILD, Kathryn Sheehan and Mary Waldner. Tulsa, OK: Council Oak Books, 1991. (Ecology)

Games, stories, activities, experiments and ideas about living lightly on planet Earth. This book is a "must have!"

EARTHSCORE, YOUR ENVIRONMENTAL AUDIT & GUIDE, Donald W. Lotter. Lafayette, CA: Morning Sun Press, 1993.

How do you score on living lightly on the planet? This book lets you know.

EARTHWATCH, EARTHCYCLES, AND ECOSYSTEMS, Beth Savan. Illustrated by Pat Cupples. Reading, PA: Addison-Wesley, 1991.

Earthwatch . . . explores how we can make the earth safe for all life, with dozens of activities.

Grades 2-7

50 SIMPLE THINGS™ KIDS CAN DO TO RECYCLE, John Javna & The Earthworks Group. California edition. Berkeley, CA: Earthworks Press, 1994.
This an "action" book.

GENERATION EARTH, Jeff Barrie. Oakland, CA: The Video Project, Media for a Safe and Sustainable World, 1995. Video, 40 minutes.
This is a look at some of the most innovative environmental education projects in schools. Laytonville's vermicomposting and garden project has a brief moment of fame in this film. Thank you, Jeff and Tyson, for your efforts to educate! *Call* 1-800-4-planet.

GRASS GAMES AND MOON RACES: CALIFORNIA INDIAN GAMES AND TOYS, Jeannine Gendar. Berkeley, CA: Heyday Books, 1995.
". . . [A]s people from tribes around the Sierra Nevada foothills and beyond greet each other, some of them tak[e] their places in two long, opposing rows. Off to the side, someone is keeping track of bets, and spectators are finding comfortable spots . . . [K]erchiefs and bones and counters come out, and the singing begins. This is 'hand game,' the classic Native American guessing game, played in California." When I took my students on a week-long field trip to Lake County, we camped at a state park to study the flora, fauna, history, and Native Americans who once lived there. This book provided the historical context of the games they still play today.
All Ages

THE GREEN TEAM: WINNING IDEAS & ACTIVITIES TO PROMOTE ENVIRONMENTAL AWARENESS, Dorothy Michener. Nashville, TN: Incentive Publications, 1993.
This is a great resource for children to see what others are doing around the state.

LEARN AND PLAY IN THE GARDEN: GAMES, CRAFTS, AND ACTIVITIES FOR CHILDREN, Meg Herd. New York, NY: Barron's Educational Series, 1995.
This is a great book for activities.

NATIVE AMERICAN GARDENING STORIES, PROJECTS AND RECIPES FOR FAMILIES, Michael J. Caduto and Joseph Bruchac. Golden, CO: Fulcrum Publishing, 1996.
This is from the authors of *Keepers of the Earth* and explores Native American gardening.

THE NO WASTE ANTHOLOGY: A TEACHER'S GUIDE TO ENVIRONMENTAL ACTIVITIES K-12. Sacramento, CA: California Department of Health Services, 1991. (Toxic Substances Control Program)
Contains lesson plans.

RECYCLE REX™ EDUCATION & RECYCLING: EDUCATOR'S WASTE MANAGEMENT RESOURCE AND ACTIVITY GUIDE 1994. Sacramento, CA: Department of Conservation, Department of Recycling, 1994.
You can get this publication from the State Department in California.

REDUCING, REUSING, RECYCLING, Bobbie Kalman. New York, NY: Crabtree Publishing, 1991. (Crabtree Environmental Series)
Shows children involved in recycling.

THE RODALE BOOK OF COMPOSTING, Deborah L. Martin and Grace Gershuny. Emmaus, PA: Rodale Press, 1992.
I often use this great book .

SAVE THE EARTH, Betty Miles. Drawings by Nelle Davis. New York, NY: Alfred A. Knopf, 1991.
An action handbook for children. "A lot of people think kids can't do very much work in the world. But I think kids might be the only chance," says Anna Brown (11 years old).

SPECIAL PLACES, SPECIAL PEOPLE, Wendy Titman. Godalming, Surrey, UK: World Wide Fund for Nature, 1994.
This book is the result of a two-year study on children and their school environment. I highly recommend this book.

VOICES FROM THE FIELDS: CHILDREN OF MIGRANT WORKERS TELL THEIR STORIES, S. Beth Atkin. Boston, MA: Little, Brown, 1993.
This book chronicles the obstacles faced by the sons and daughters of migrant workers and celebrates their ability to triumph over harsh circumstances. This is a very powerful book.

WORMS EAT MY GARBAGE, Mary Appelhof. Illustrated by Mary Frances Fenton. Revised edition. Kalamazoo, MI: Flower Press, 1997.
How to set up and maintain a home vermicomposting unit. To supplement use of this textbook in my 6th-8th middle-school classes, I developed the comprehensive questions included in Appendix F.

WORMS EAT OUR GARBAGE, Mary Appelhof, Mary Frances Fenton, and Barbara Loss Harris. Kalamazoo, MI: Flower Press, 1993.
A unique curriculum for home, school, and outdoor centers that treats earthworms in an exciting and informative manner.
Grades 4-8

Bibliography

Appelhof, Mary. ***Worms eat my garbage***. Kalamazoo, MI: Flower Press, 1982. Second edition, 1997.

Appelhof, Mary, Mary Frances Fenton, and Barbara Loss Harris. ***Worms eat our garbage: classroom activities for a better environment***. Kalamazoo, MI: Flower Press, 1993.

Bear, Firman E. ***Earth: the stuff of life***, second edition, revised by H. Wayne Pritchard and Wallace E. Akin. Norman, OK: University of Oklahoma Press, 1986.

Bear, Firman E. ***Soils & fertilizer***, third edition. NY: Wiley & Sons, 1942.

Begley, Sharon, and Mary Hager. "Pesticides and kids' risks," ***Newsweek***, 130(22):1998 June 1; p56-57.

California Department of Conservation. ***Recycle rex education and recycling: educator's waste management resource and activity guide 1994.*** Sacramento, CA: California Department of Conservation, Division of Recycling, 1994.

Dromgoole, John. ***Catalog***, vol. 6. Austin, TX: The Natural Gardener, 1995.

Edwards, C. A., and J.R. Lofty. ***Biology of earthworms***, second edition. London, UK: Chapman & Hall, 1977.

Francis, Robert. ***The complete book of compost***. New York, NY: Berkeley Publishing Group, 1998.

Golueke, Clarence G. ***Composting: a study of the process and its principles***. Emmaus PA: Rodale Press, 1975.

Goodlad, John I. ***A place called school***. New York, NY: McGraw-Hill, 1984.

Hartenstein, Roy, E.F. Neuhauser, and D.L. Kaplan. "Reproductive potential of the earthworm *Eisenia foetida*," ***Oecologia***, 43:1979; p329-340.

Jaffe, Roberta, and Gary Appel. ***Growing classroom garden-based science***, illustrated by Kate Murry. Menlo Park, CA: Addison-Wesley, 1990.

Javna, John, and the Earthworks Group. ***50 simple things kids can do to recycle,*** California edition. Berkeley CA: Earthworks Press, 1994.

Kalman, Bobbie. ***Reducing, reusing, recycling***. New York, NY: Crabtree Publishing, 1991.

Klemm, Donald J. ***A guide to the freshwater Annelida.*** Dubuque, IA: Kendall/Hunt Publishing, 1985.

Koeph, Herbert H., Bo D. Pettersson, and Wolfgang Schaumann. ***Bio-dynamic agriculture: an introduction***. Spring Valley, NY: Anthroposophic Press, 1976.

Lotter, Donald W. ***EarthScore: your environmental audit and guide***. Lafayette, CA: Morning Sun Press, 1993.

Martin, Deborah L, and Grace Gershuny. ***The Rodale book of composting***. Emmaus, PA: Rodale Press, 1992.

Marzano, Robert. "Beyond buzzwords," ***Learning***, 25:1997 Jan/Feb.

Michener, Dorothy. ***The green team: winning ideas and activities to promote environ mental awareness***. Nashville, TN: Incentive Publications, 1993.

Miles, Betty. ***Save the earth: an action handbook for kids***. New York, NY: Alfred A. Knopf, 1991.

Miller, B. Tyler, Jr. ***Living in the environment***, sixth edition. Belmont, CA: Wadsworth Publishing, 1990.

National Academy of Sciences. ***Agricultural alternatives***. Washington, DC: Office of Technology Assessment, 1989.

Organic Gardening Magazine Staff. ***The encyclopedia of organic gardening,*** new revised edition. Emmaus, PA: Rodale Press, 1978.

Orr, David W. ***Earth in mind: on education, environment and the human prospect.*** Washington, DC: Island Press, 1991.

Perry, Ted. "A letter to all the people," attributed to Chief Seattle. Hollywood, CA: Southern Baptist Convention, 1977.

"Questions Editor," ***Garbage magazine*** (1992 May/June), p66.

Reinecke, Adrian. Personal communication to Mary Appelhof, 1996 July.

Roulac, John W. **Backyard composting: your complete guide to recycling yard clippings**. Ojai, CA: Harmonious Press, 1992.

Seattle, Chief (1854). *See* Perry, Ted.

Shapiro, Laura. "Is organic better?" ***Newsweek***, 130(22):1998 June 1; p54-57.

Share, Elizabeth, and Laurette Rogers. "Get real, project based learning," ***Learning***, 25: 1997 Jan/Feb; p64-65.

Sheehan, Kathryn, and Mary Waidner. ***Earth child: games, stories, activities, experiments and ideas about living lightly on the planet Earth***. Tulsa, OK: Council Oaks Books, 1991.

Simon, Seymour. ***What do you want to know about earthworms?*** New York, NY: Scholastic Book Services, 1969.

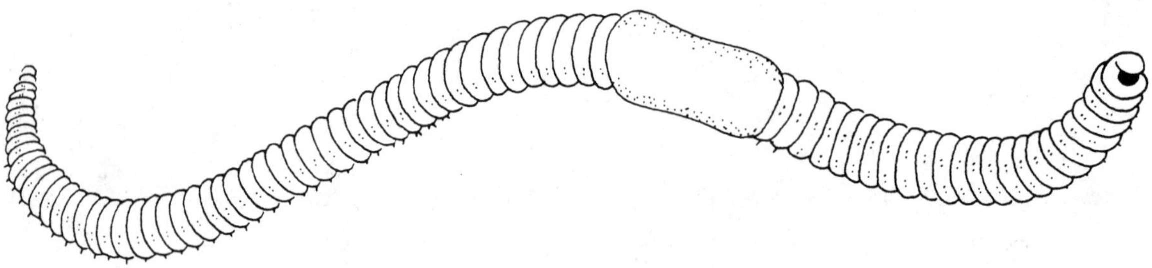

Index

Page numbers in **bold** refer to definitions; those in *italics* refer to figures.

About the Author

Binet Payne was born in Whittier, California. In the late 1960s, searching for the simpler rural life she experienced as a child in southern California, she moved north to Mendocino County. She and her husband purchased 14 acres and built a home in the small logging community of Laytonville, where wild creatures really do "go bump in the night." Among her most cherished moments are reading to her three granddaughters, Ashley, Serena, and Caitlyn. Presently she serves as teacher, consultant, project director, and grant writer.

A graduate of Sonoma State University, Binet received her teaching credentials from Dominican College of San Rafael. Her experience began with children in preschool and then extended through the ninth grade where gardening has been a part of their education since 1986. Believing that rural education needs to be preserved, her extremely effective teaching connects a sense of place to the intellectual work students do. She has worked with the Center for Ecoliteracy, the Autodesk Foundation, and the Center for Complex Instruction at Stanford University.

Author of numerous articles in periodicals, magazines, and newspapers, Binet has developed nationally recognized programs to teach ecology. Her principles of ecoliteracy and project-based learning for all students are the basis for local and national networks of schools and businesses to promote educational gardens, recycling, and vermicomposting programs.

Currently, Binet is a project director for the North Coast Rural Challenge Network, helping students become stewards of their communities rather than irresponsible owners. NCRCN supports schools as powerful learning centers for the whole community. Education integrated with community issues offers a rich and powerful environmental context for learning. "In Laytonville, we do not want our children to be our chief export."

Book designer: Nancy Essex
Fonts: Benquait, Palatino, Helvetica, Dom Casual
Editor, indexer: WordenDex Plus,
Kalamazoo, MI
Proofreader: Maria Perez-Stable
Cover layout: Fritz Design and Consulting,
Portage, MI
Printer: Malloy Lithographic, Inc.,
Ann Arbor, MI

Flower Press